Solving Statics Problems in MATLAB®

Brian D. Harper
Ohio State University

ENGINEERING MECHANICS
STATICS

Sixth Edition

J. L. Meriam
L. G. Kraige
*Virginia Polytechnic Institute
and State University*

John Wiley & Sons, Inc.

MATHLAB is a registered trademark of The MathWorks, Inc.

Cover Photo: © Medioimages/Media Bakery

To order books or for customer service please, call 1-800-CALL WILEY (225-5945).

ISBN-13   978- 0-470-09925-4
ISBN-10   0-470-09925-9

10 9 8 7 6 5 4 3 2 1

CONTENTS

INTRODUCTION

Computers and software have had a tremendous impact upon engineering education over the past several years and most engineering schools now incorporate computational software such as MATLAB in their curriculum. Since you have this supplement the chances are pretty good that you are already aware of this and will be having to learn to use MATLAB as part of a Statics course. The purpose of this supplement is to help you do just that.

There seems to be some disagreement among engineering educators regarding how computers should be used in an engineering course such as Statics. I will use this as an opportunity to give my own philosophy along with a little advice. In trying to master the fundamentals of Statics there is no substitute for hard work. The old fashioned taking of pencil to paper, drawing free body diagrams and struggling with equilibrium equations, etc. is still essential to grasping the fundamentals of mechanics. A sophisticated computational program is not going to help you to understand the fundamentals. For this reason, my advice is to use the computer only when required to do so. Most of your homework can and should be done without a computer. A possible exception might be using symbolic algebra to check some messy calculations.

The problems in this booklet are based upon problems taken from your text. The problems are slightly modified since most of the problems in your book do not require a computer for the reasons discussed in the last paragraph. One of the most important uses of the computer in studying Mechanics is the convenience and relative simplicity of conducting parametric studies. A parametric study seeks to understand the effect of one or more variables (parameters) upon a general solution. This is in contrast to a typical homework problem where you generally want to find one solution to a problem under some specified conditions. For example, in a typical homework problem you might be asked to find the reactions at the supports of a structure with a concentrated force of magnitude 200 lb that is oriented at an angle of 30 degrees from the horizontal. In a parametric study of the same problem you might typically find the reactions as a function of two parameters, the magnitude of the force and its orientation. You might then be asked to plot the reactions as a function of the magnitude of the force for several different orientations. A plot of this type is very beneficial in visualizing the general solution to a problem over a broad range of variables as opposed to a single case.

As you will see, it is not uncommon to find Mechanics problems that yield equations that cannot be solved exactly. These problems require a numerical approach that is greatly simplified by computational software such as MATLAB. Although numerical solutions are extremely easy to obtain in MATLAB this is still the method of last resort. Chapter 1 will illustrate several methods for obtaining symbolic (exact) solutions to problems. These methods should always be tried first. Only when these fail should you generate a numerical approximation.

Many students encounter some difficulties the first time they try to use a computer as an aid to solving a problem. In many cases they are expecting that they have to do something fundamentally different. It is very important to understand that there is no fundamental difference in the way that you would formulate computer problems as opposed to a regular homework problem. Each problem in this booklet has a problem formulation section prior to the solution. As you work through the problems be sure to note that there is nothing peculiar about the way the problems are formulated. You will see free-body diagrams, equilibrium equations etc. just like you would normally write. The main difference is that most of the problems will be parametric studies as discussed above. In a parametric study you will have at least one and possibly more parameters or variables that are left undefined during the formulation. For example, you might have a general angle θ as opposed to a specific angle of $20°$. If it helps, you can "pretend" that the variable is some specific number while you are formulating a problem.

This supplement has seven chapters. The first chapter contains a brief introduction to MATLAB. If you already have some familiarity with MATLAB you can skip this chapter. Although the first chapter is relatively brief it does introduce all the methods that will be used later in the book and assumes no prior knowledge of MATLAB. Chapters 2 through 7 contain computer problems taken from chapters 2 through 7 of your textbook. Thus, if you would like to see some computer problems related to friction you can look at the problems in chapter 6 of this supplement. Each chapter will have a short introduction that summarizes the types of problems and computational methods used. This would be the ideal place to look if you are interested in finding examples of how to use specific functions, operations etc.

This supplement uses the student edition of MATLAB version 7.1. MATLAB is a registered trademark of The Mathworks, Inc., 24 Prime Park Way, Natick, Massachusetts, 01760.

AN INTRODUCTION
TO MATLAB

1

This chapter provides an introduction to the MATLAB programming language. Although the chapter is introductory in nature it will cover everything needed to solve the computer problems in this booklet.

1.1 Numerical Calculations

When you open MATLAB you should see a prompt something like the following.

EDU»

To the right of this prompt you will write some sort of command. The following example assigns a value of 200 to the variable a. Simply type "a = 200" at the prompt and then press enter.

EDU» a=200
a =
   200

There are many situations where you may not want MATLAB to print the result of a command. To suppress the output simply type a semi-colon ";" after the command. For example, to suppress the output in the above case type "a = 200;" and the press enter. The basic mathematical operations of addition, subtraction, multiplication, division and raising to a power are accomplished exactly as you would expect if you are familiar with other programming languages, i.e. with the keys +, -, *, \, and ^. Here are a couple of examples.

EDU» c=2*5-120/4
c =
  -20

EDU» d=2^4
d =
   16

MATLAB has many built in functions that you can use in calculations. If you already know the name of the function you can simply type it in. If not, you can find a list of functions in the "Help Desk (HTML)" file available under Help in the main menu bar. Here are a few examples.

EDU» e=sin(.5)+tan(.2)
e =
   0.6821

As with most mathematical software packages, the default unit for angles will be radians.

EDU» f=sqrt(16)
f =
    4

EDU» g=log(5)
g =
   1.6094

EDU» h=log10(5)
h =
   0.6990

In the last two examples be careful to note that *log* is the natural logarithm. For a base 10 logarithm (commonly used in engineering) you need to use *log10*.

Range Variables

In the above examples we have seen many cases where a variable (or name) has been assigned a numerical value. There are many instances where we would like a single variable to take on a range of values rather than just a single value. Variables of this type are often referred to as range variables. For example, to have the variable x take on values between 0 and 3 with an increment of 0.25 we would type "x=0:0.25:3".

EDU» x=0:0.25:3
x =
 Columns 1 through 7
      0   0.2500   0.5000   0.7500   1.0000   1.2500   1.5000
 Columns 8 through 13

   1.7500   2.0000   2.2500   2.5000   2.7500   3.0000

Range variables are very convenient in that they allow us to evaluate an expression over a range with a single command. This is essentially equivalent to performing a loop without actually having to set up a loop structure. Here's an example.

```
EDU» x=0:0.25:3;
EDU» f=3*sin(x)-2*cos(x)
f =
  Columns 1 through 7
  -2.0000  -1.1956  -0.3169  0.5815  1.4438  2.2163  2.8510
  Columns 8 through 13
   3.3084   3.5602   3.5906  3.3977  2.9936  2.4033
```

Although range variables are very convenient and almost indispensable when making plots, they can also lead to considerable confusion for those new to MATLAB. To illustrate, suppose we wanted to compute a value $y = 2*x-3*x^2$ over a range of values of x.

```
EDU» x=0:0.5:3;
EDU» y = 2*x-3*x^2
??? Error using ==> mpower
Matrix must be square.
```

What went wrong? We have referred to x as a range variable because this term is most descriptive of how we will actually use variables of this type in this manual. As the name implies, MATLAB is a program for doing matrix calculations. Thus, internally, x is really a row matrix and operations such as multiplication or division are taken, by default, to be matrix operations. Thus, MATLAB takes the operation x^2 to be matrix multiplication x*x. Since matrix multiplication of two row matrices doesn't make sense, we get an error message. In a certain sense, getting this error message is very fortunate since we are not actually interested in a matrix calculation here. What we want is to calculate a certain value y for each x in the range specified. This type of operation is referred to as term-by-term. For term-by-term operations we need to place a period "." in front of the operator. Thus, for term by term multiplication, division, and raising to a power we would write ".*", "./" and ".^" respectively. Since matrix and term-by-term addition or subtraction are equivalent, you do not need a period before + or -. The reason that our earlier calculations did not produce an error is that they all involved scalar quantities. To correct the error above we write.

```
EDU» x=0:0.5:3;
EDU» y = 2*x-3*x.^2
y =
```

    0   0.2500 -1.0000 -3.7500 -8.0000 -13.7500 -21.0000

Note that we wrote "2*x" instead of "2.*x" since 2 is a scalar. If in doubt, it never hurts to include the period.

1.2 Writing Scripts (m-files)

If you have been working through the examples above you have probably noticed by now that you cannot modify a line once it has been entered into the worksheet. This makes it rather difficult to debug a program or run it again with a different set of parameters. For this reason, most of your work in MATLAB will involve writing short scripts or m-files. The term m-file derives from the fact that the programs are saved with a ".m" extension. An m-file is essentially just a list of commands that you would normally enter at a prompt in the worksheet. After saving the file you just enter the name of the file at the prompt after which MATLAB will execute all the lines in the file as if they had been entered. The chief advantage is that you can alter the script and run it as many times as you like.

To get started, select *"File...New...M-File"* from the main menu. A screen like that below will then pop up.

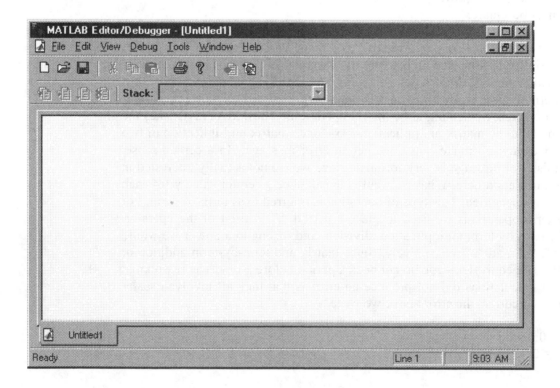

Now enter the program line by line into the editor. Basically, you just enter the commands you would normally enter at a prompt (do not, however, enter >> at the beginning of a line). If you want to enter comments, begin the line with "%". These lines will not be executed. Once you have finished entering the script into the editor you can save the file (be sure to remember where you saved it). The file name can contain numbers but cannot begin with a number. It also cannot contain any mathematical operators such as +, -, *, / etc. Also, the name of the file should not be the same as a variable it computes.

To run the script all you have to do is type the name of the program at a prompt (>>), but without the ".m" extension. A common error at this point is that MATLAB will give some response indicating that it hasn't a clue what you are doing. The reason this happens is that the file you saved is not in the current path. You can check (or change) the path by selecting *File...Set Path...* from the main menu.

Following is a simple example of a script file that sets up a range variable x and then calculates two functions over the range.

```
%%%%%%%%%%%%%%%%%%%%%%%%%%%%%%%%%%%%%%%%%%%%%
% This script calculates two
% functions over a specified range
x=0:0.2:1;
f=sin(x)./(1+x)
g=sin(x).*exp(x)
%%%%%%%%% end of script %%%%%%%%%%%%%%%%%%%%%
```

The program was saved as example1, so here is the output you should see when the program is entered at a prompt.

```
EDU» example1
f =
     0   0.1656   0.2782   0.3529   0.3985   0.4207
g =
     0   0.2427   0.5809   1.0288   1.5965   2.2874
```

To save space we will, in the future, show the output of a script immediately after the script is given without actually showing the file name being entered at a prompt.

1.3 Defining Functions

Although MATLAB has many built in functions (see the "Help Desk (HTML)" file available under Help in the main menu bar), it is sometimes advantageous to define your own functions. While user defined functions are actually special types of m-files (they will be saved with the .m extension) their use is really quite different from the m-files discussed above. For example, suppose you were to create a function that you call f1. You would then use this function as you would any built in function like sin, cos, log etc.

Since functions are really special types of m-files you start by opening an editor window with "*File...New...M-File*". The structure of the file is, however, quite different from a script file. Following is an m-file for a function called load1. Note that you should save the file under the same name as the function. The file would thus show up as load1.m in whatever directory you save it. As with script files, you need to be sure the current path contains the directory in which the file was stored.

```
%%%%% a function m-file %%%%%%%%%%%%%%%%
function y = load1(x)
% this file defines the function load1(x)
y = 2*(x-1)+3*(x-2).^2;
%%%%% end of function m-file %%%%%%%%%%
```

It is important to understand the structure of a function definition before you try to create some on your own. The name of this function is "load1" and will be used as you would any other function. y and x are local variables which MATLAB uses to calculate the function. There is nothing special about y and x, other variable names will work as well. Now, suppose you were to type load1(3) at a prompt or within a script file. MATLAB will calculate the expression for y substituting x = 3. This value of y will then be returned as the value for load1(3).

Here are a few examples using our function load1.

EDU» load1(3)
ans =
   7

EDU» f=load1(5)-load1(3)
f =
   28

EDU» load1(load1(3))
ans =
   87

EDU» x=1:0.2:2;
EDU» g=load1(x)
g =
   3.0000   2.3200   1.8800   1.6800   1.7200   2.0000

1.4 Graphics

One of the most useful things about a computational software package such as MATLAB is the ability to easily create graphs of functions. As we will see, these graphs allow one to gain a lot of insight into a problem by observing how a solution changes as some parameter (the magnitude of a load, an angle, a dimension etc.) is varied. This is so important that practically every problem in this supplement will contain at least one plot. By the time you have finished reading this supplement you should be very proficient at plotting in MATLAB. This section will introduce you to the basics of plotting in MATLAB.

MATLAB has the capability of creating a number of different types of graphs. Here we will consider only the X-Y plot. The most common and easiest way to generate a plot of a function is to use range variables. The following example will guide you through the basic procedure.

EDU» x=-3:0.1:3;
EDU» f=x.*exp(-x.^2);
EDU» plot(x,f)

The procedure is very simple. First define a range variable covering the range of the plot, then define the function to be plotted and issue the plot command. After typing in the above you should see a graph window pop up which looks something like the following figure.

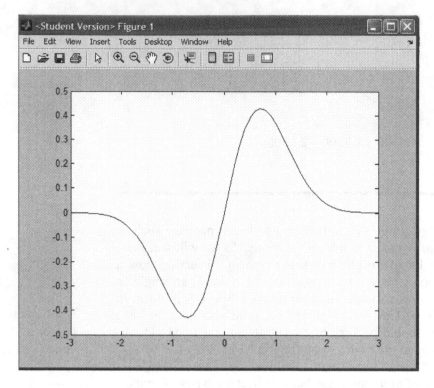

Things like titles and axis labels can be added in one of two ways. The first is by line commands such as those shown below.

EDU» xlabel('x')
EDU» ylabel('x*exp(-x^2)')
EDU» title('A Simple Plot')

Take a look at the graph window after you type each of the lines above. At this point, the graph window should look something like that shown below.

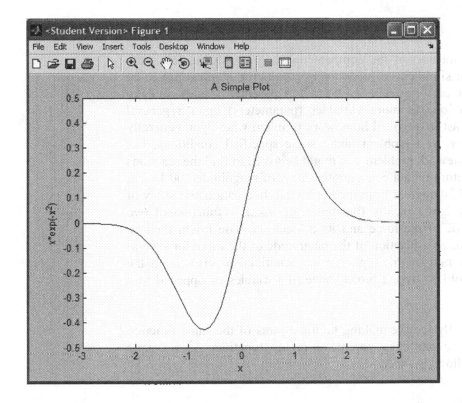

There are many other things that you can add or change on the plot. The easiest way to do this is with the second of the two approaches mentioned above. In the graph menu select "*Tools...Edit Plot*". Now you can edit the plot in a number of different ways. To see some of the options try either double clicking or right clicking on the graph. You should spend some time experimenting with a graph window to see the range of possibilities. In particular, be sure to try clicking the *show plot tools button*.

Once you have things like you want them you can save the graph to a file. You can also export the graph to some format suitable for inserting in, say, a word processor. Select "*File...Export Setup*" in the graph window to see the range of possible formats.

Parametric Studies

One of the most important uses of the computer in studying Mechanics is the convenience and relative simplicity of conducting parametric studies (not to be confused with parametric plotting discussed below). A parametric study seeks to understand the effect of one or more variables (parameters) upon a general solution. This is in contrast to a typical homework problem where you generally want to find one solution to a problem under some specified conditions. For example, in a typical homework problem you might be asked to find the reactions at the supports of a structure with a concentrated force of magnitude 200 lb that is oriented at an angle of 30 degrees from the horizontal. In a parametric study of the same problem you might typically find the reactions as a function of two parameters, the magnitude of the force and its orientation. You might then be asked to plot the reactions as a function of the magnitude of the force for several different orientations. A plot of this type is very beneficial in visualizing the general solution to a problem over a broad range of variables as opposed to a single case.

Parametric studies generally require making multiple plots of the same function with different values of a particular parameter in the function. As a simple example, consider the following function.

$$f = 5 + x - 5x^2 + ax^3$$

What we would like to do is gain some understanding of how f varies with both x and a. It might be tempting to make a three dimensional plot in a case like this. Such a plot can, in some cases, be very useful. Usually, however, it is too difficult to interpret. This is illustrated by the following three dimensional plot of f versus x and a.

```
% This script produces a 3-d plot of
% f versus x and a.
x=linspace(-10,10,30);
a=linspace(-3,3,30);
[X,A]=meshgrid(x,a);
F=5+X-5*X.^2+A.*X.^3;
B=0.*A+2;
mesh(X,A,F)
colormap gray
%%%%%%% end of script %%%%%%%%%%%%%%%%%%%%
```

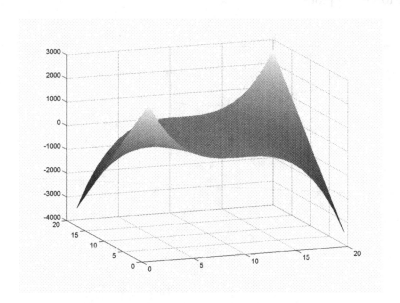

The plot above certainly is interesting but, as mentioned above, not very easy to interpret. In most cases it is much better to plot the function several times (with different values of the parameter of interest) on a single two-dimensional graph. We will illustrate this by plotting f as a function of x for a = -1, 0, and 1. This is accomplished in the following script.

```
% script for plotting f versus x for several
% values of a.
x=-5:0.1:5;
g=5+x-5*x.^2;
f1=g-x.^3; % a = -1
f2=g;       % a = 0
f3=g+x.^3; % a = 1
plot(x,f1,x,f2,x,f3)
xlabel('x')
ylabel('f')
%%%%%%% end of script %%%%%%%%%%%%%%%%%%%%%%
```

Running this script results in the following plot.

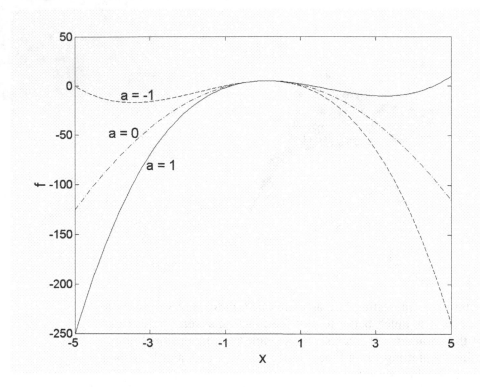

Parametric Plots

It often happens that one needs to plot some function y versus x but y is not known explicitly as a function of x. For example, suppose you know the x and y coordinates of a particle as a function of time but want to plot the trajectory of the particle, i.e. you want to plot the y coordinate of the particle versus the x coordinate. A plot of this type is generally called a parametric plot. Parametric plots are easy to obtain in MATLAB. You start by defining the two functions in terms of the common parameter and then define the common parameter as a range variable. Then issue a plot command using the two functions as arguments. The following script illustrates this procedure.

```
% This script file illustrates parametric plotting.
% A function f is plotted versus another function g.
% The two functions are related by the
% common parameter a.
a=-1:0.05:3.5;
f=10*a.*(2-a);
g=sin(3*a);
plot(g,f)
xlabel('g')
ylabel('f')
%%%%%%% end of script %%%%%%%%%%%%%%%%%%%%%%%%%%%%%
```

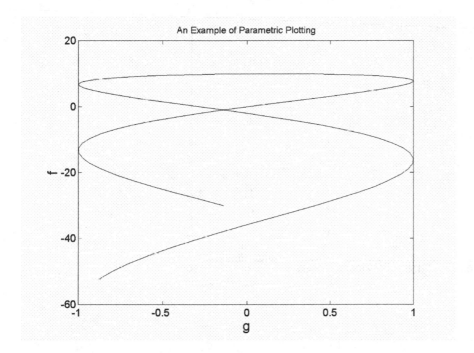

The range selected for the parameter can have a big, and sometimes surprising effect on the resulting graph. To illustrate, try increasing the upper limit on the range on a a few times and see how the graph changes.

You can, of course, also plot g as a function of f.

```
% Another example of parametric plotting
a=-1:0.05:6;
f=10*a.*(2-a);
g=sin(3*a);
plot(f,g)
xlabel('f')
ylabel('g')
%%%%%%%%% end of script %%%%%%%%%%%%%%%%%%
```

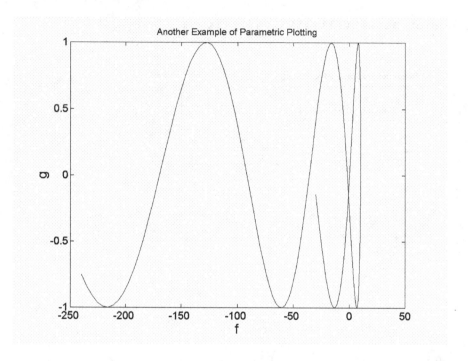

1.5 Symbolic Calculations

One of the most useful features of many modern mathematical software packages is the capability of doing symbolic math. By symbolic math we mean mathematical operations on expressions containing undefined variables rather than numbers. Maple is, perhaps, the best known computer algebra program. Many other programs (including MATLAB) have a Maple engine that is called whenever symbolic operations are performed. The Maple engine is part of the Symbolic Math Toolbox so, if you do not have this toolbox, you will not be able to do symbolic math. If you have a student edition of MATLAB you should have the Symbolic Math Toolbox. If you do not have the Symbolic Math Toolbox, you will need to perform whatever symbolic calculations are done in this manual by hand.

How does MATLAB decide whether to call Maple and attempt a symbolic calculation? Perhaps the best way to learn this is by studying the examples below and by practicing. Maple will always be used to manipulate any expression represented as a character string (i.e. an expression enclosed in 'quotation' marks). In the remainder of this manual we will refer to character strings of this type as Maple expressions. Here are a few examples of Maple expressions.

```
EDU» 'x+a*x^2-b*x^3'
ans =
x+a*x^2-b*x^3
EDU» f='A*sin(x)*exp(-x^2)'
f =
A*sin(x)*exp(-x^2)
```

Instead of placing the entire expression in quotations, you can also first declare certain variables to be symbols and then write the expressions as usual. This is more convenient in some situations but can get confusing. Here are some examples of this type.

```
EDU» syms x a b A          % Sets x, a, b and A as symbols
EDU» g=x+a*x^2-b*x^3
g =
x+a*x^2-b*x^3
EDU» f=A*sin(x)*exp(-x^2)
f =
A*sin(x)*exp(-x^2)
```

The above are just examples of Maple expressions and do not involve any symbolic math. Two of the most important applications of symbolic math will be discussed in the next two sections, namely symbolic calculus (integration and differentiation) and symbolic solution of one or more equations. The purpose of the present section is to introduce you to the basic procedures of symbolic math as well as to give a few other useful applications. One particularly useful symbolic operation is substitution and is illustrated by the following examples.

```
EDU» g='x+a*x^2-b*x^3'
g =
x+a*x^2-b*x^3

EDU» subs(g,'a',3)    % substitutes 3 for a in g. Note the quotation marks about a.
ans =
x+3*x^2-b*x^3

EDU» g1=subs(g,'x','(y-c)')
g1 =
((y-c))+a*((y-c))^2-b*((y-c))^3

EDU» syms x y A B
EDU» f=A*sin(x)*exp(-B*x^2)
f =
A*sin(x)*exp(-B*x^2)

EDU» subs(f,B,4)   % Note that B doesn't have to be in quotes since it
                             was declared to be a symbol.
ans =
A*sin(x)*exp(-4*x^2)
```

Another very convenient feature available in the Symbolic Toolbox provides the capability of defining "inline" functions thus avoiding the necessity for creating special m-files as described above. Here is an example.

```
EDU» f=inline('x^2*exp(-x)')
f =
    Inline function:
    f(x) = x^2*exp(-x)

EDU» f(3)
ans =
    0.4481
```

In the previous section we plotted the function $f = 5 + x - 5x^2 + ax^3$ for several values of a. It turns out to be somewhat more convenient to do this with an inline function. Let's see how it works.

EDU» f=inline('5+x-5*x^2+a*x^3')
f =
    Inline function:
    f(a,x) = 5+x-5*x^2+a*x^3

At this point we might anticipate a minor problem. When we plot this as a function of x we will have a range variable as one of the arguments for f. This creates a problem as MATLAB will be performing the numerical calculations and it will take the operations in f as matrix operations rather than term by term. Fortunately we can use the *"vectorize"* operator to insert periods in the appropriate places.

EDU» f=vectorize(f)
f =
    Inline function:
    f(a,x) = 5+x-5.*x.^2+a.*x.^3

EDU» x=-5:0.1:5;
EDU» plot(x,f(-1,x),x,f(0,x),x,f(1,x))

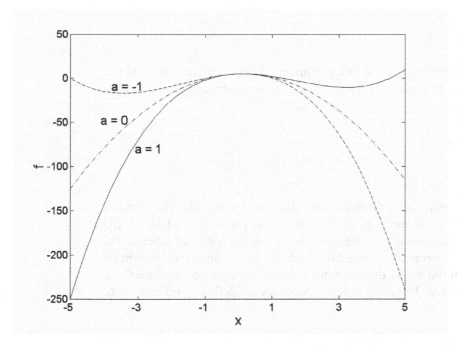

1.6 Differentiation and Integration

Here we will consider only symbolic differentiation and integration since these are most useful in the mechanics applications in this manual. For differentiation we will use the Maple *diff* command. Care should be exercised here since *diff* is also a MATLAB command. To guarantee that you are using the Maple command, be sure that you use *diff* only on Maple expressions.

```
EDU» f='a*sec(b*t)';
EDU» diff(f, 't')      % Note that t must be placed in quotation marks since
                          it is a symbol.
ans =
a*sec(b*t)*tan(b*t)*b
```

We can avoid using the quotation marks by making a, b and t symbols.

```
EDU» syms a b t
EDU» f=a*sec(b*t);
EDU» diff(f, t)
ans =
a*sec(b*t)*tan(b*t)*b
```

To find higher order derivatives use the command diff(f, x, n) where n is the order of the derivative. For example, to find the third derivative of a*log*(b+x) we would write

```
EDU» diff('a*log(b+x)', 'x', 3)
ans =
2*a/(b+x)^3
```

Symbolic integration will be performed with the *int* command. The general format of this command is int(f, x, a, b) where f is the integrand (a Maple expression), x is the integration variable and a and b are the limits of integration. If x, a and b have not been declared symbols with the *syms* command, they must be placed in quotation marks. If the integration limits are omitted, the indefinite integral will be evaluated. Here are several examples of definite and indefinite integrals.

```
EDU» int('sin(b*x)','x')
ans =
-1/b*cos(b*x)
```

```
EDU» int('sin(b*x)','x','c','d')
ans =
 -(cos(b*d)-cos(b*c))/b
```

```
EDU» syms x b c d
EDU» f=b*log(x);
EDU» int(f,x)
ans =
b*x*log(x)-b*x
```

```
EDU» int(f,x,c,d)
ans =
b*d*log(d)-b*d-b*c*log(c)+b*c
```

If a definite integral contains no unknown parameters either in the integrand or the integration limits, the *int* command will provide numerical answers. Here are a few examples.

```
EDU»  int('x+3*x^3','x',0,3)
ans =
261/4
```

```
EDU» int('log(x)','x',2,5)   % Don't forget that log is the natural logarithm.
ans =
5*log(5)-3-2*log(2)
```

Note that Maple will always try to return an exact answer. This usually results in answers containing fractions or functions as in the above examples. This is very useful in some situations; however, one often wants to know the numerical answer without having to evaluate a result such as the above with a calculator. To obtain numerical answers use Maple's *eval* function as in the following examples.

```
EDU» eval(int('x+3*x^3','x',0,3))
ans =
  65.2500
EDU» eval(int('log(x)','x',2,5))
ans =
  3.6609
```

1.7 Solving Equations

Our preferred approach to solving equations is to use Maple's *solve* command available in the *Symbolic Toolbox*. We will also discuss briefly MATLAB's built in function *fzero* which can be used to solve single equations numerically.

Solving single equations

Here are a couple of examples illustrating how to use Maple's *solve* command to solve a single equation symbolically. The first example is the (hopefully) familiar quadratic equation.

```
EDU» f='a*x^2+b*x+c=0';
EDU» solve(f, 'x')
ans =
 1/2/a*(-b+(b^2-4*a*c)^(1/2))
 1/2/a*(-b-(b^2-4*a*c)^(1/2))


EDU» g='a*sin(x)-cos(x)=1';
EDU» solve(g, 'x')
ans =
                              pi
 atan2(2*a/(1+a^2),(a^2-1)/(1+a^2))
```

Note that both examples produced two solutions with each solution being placed on separate lines. One of the solutions to the second example is the number π. The second solution illustrates something you should always watch out for. In many cases you may get a solution in terms of a function that you are not familiar with, i.e. atan2 that appears in the second solution to the second example above. Whenever this happens you should type "help atan2" at a prompt in order to learn more about the function.

If the variable being solved for is the only unknown in the equation, *solve* will return a number as the result. Here are a couple of examples.

```
EDU» f='2*x^2+4*x-12';
EDU» solve(f, 'x')
ans =
 -1+7^(1/2)
 -1-7^(1/2)


EDU» g='5*sin(x)-cos(x)=1';
EDU» solve(g, 'x')
ans =
```

pi
atan(5/12)

If you want the answer as a floating point number instead of an exact result, use the Maple command *eval*. For example, with g defined as above we would write

EDU» eval(solve(g, 'x'))
ans =
   3.1416
   0.3948

Although Maple will always attempt to return an exact (symbolic) result, you may occasionally have a case where a floating point number is returned without using *eval*. Here is an example.

EDU» f='10*sin(z)=2*z+1';
EDU» solve(f, 'z')
ans =
.12541060097666150276238831440340

Whenever this occurs, Maple was unable to solve the equation exactly and switched automatically to a numerical solution. In a certain sense this is very convenient as the same command is being used for both symbolic and numerical calculations. The problem is that nonlinear equations quite often have more than one solution, as illustrated by the examples above. If you have the full version of Maple then you will have various options available for finding multiple numerical solutions. But since we are dealing now with numerical rather than symbolic solutions it makes more sense just to use MATLAB.

The appropriate MATLAB command for the numerical solution of an equation is *fzero*. This command numerically determines the point at which some defined function has a value of zero. This means that we need to rewrite our equation in terms of an expression whose zeros (roots) are solutions to the original expression. This is easily accomplished by rearranging the original equation into the form *expression* = 0. We will call this expression g in this example. Rewriting the above equation f we have $g = 2z + 1 - 10\sin(z)$. The general format for fzero is fzero('function_name', x_0) where 'function_name' is the name of the m-file function that you have created and x_0 is an initial guess at the root (zero) of the function. The initial guess is very important since, if there is more than one solution, MATLAB will find the solution closest to the initial guess. Well, you may be wondering how we can determine an approximate solution if we do not yet know the solution. Actually, this is very easy to do. First we define a function g(z) whose roots will be the solution to the equation of interest. Next, we plot this function in order to estimate the location of points where g(z) = 0. Here's how it works for the present example.

```
%%%%% function m-file %%%%%%
function y = gz(z)
y = 2*z + 1 - 10*sin(z);
%%%%% end of m-file %%%%%%%%
```

Above is our user defined function (m-file) which we have named gz. Now we can plot this function over some range as illustrated below.

EDU» z=-1:0.01:3;
EDU» plot(z,gz(z),z,0) % we also plot a line at g=0 to help find the roots.
EDU» xlabel('z')
EDU» ylabel('g(z)')

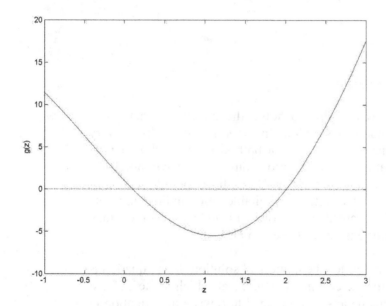

The plot shows that the function is zero at about z = 0 and 2. The first root is clearly that found by Maple's *solve* command above. Now we can easily find both roots with the *fzero* command.

EDU» fzero('gz', 0)
ans =
   0.1254

EDU» fzero('gz', 2)
ans =
   2.4985

Thus, the two solutions are z = 0.1254 and 2.4985.

Finding Maxima and Minima of Functions

Here we will illustrate two methods for finding maxima and minima of a function.

Method 1. Using the symbolic diff and solve functions.

The usual method for finding maxima or minima of a function f(x) is to first determine the location(s) x at which maxima or minima occur by solving the equation $\dfrac{df}{dx} = 0$ for x. One then substitutes the value(s) of x thus determined into f(x) to find the maximum or minimum. Consider finding the maximum and minimum values of the following function:

$$f = 1 + 2x - x^3$$

Before proceeding, it is a good idea to make a plot of the function f. This will give us a rough idea where the minima and maxima are as a check on the results we obtain below. The following script will plot f as a function of x.

```
% This script plots f as a function of x
x = -2:0.05:2;
f = 1 + 2*x - x.^3;
plot(x, f)
xlabel('x')
ylabel('f')
%%%%%%%%% end of script %%%%%%%%%%%%%
```

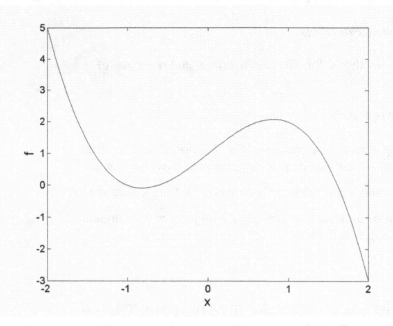

This plot shows we have a minimum of about 0 at x of about −0.9 and a maximum of about 2 at x of about 0.9. Now we can proceed to find the results more precisely.

EDU» f = '1+2*x-x^3'

f =
1+2*x-x^3

EDU» dfdx = diff(f,'x')
 dfdx =
 2-3*x^2

EDU» solve(dfdx,'x')
 ans =
  1/3*6^(1/2)
-1/3*6^(1/2)

If an equals sign is omitted (as above) *solve* will find the root of the supplied Maple expression. In other words, it will solve the equation dfdx = 0. It is more convenient in the present case to have the solutions as floating point numbers. Thus, we should have used *eval* as in the following.

```
EDU» eval(solve(dfdx,'x'))
ans =
 0.8165
-0.8165
```

The above results are the locations (x) where the minima and maxima occur. To find the maximum and minimum values of the function f we use subs to substitute these results back into f.

```
EDU» subs(f,'x',0.8165)
ans =
2.0887
```

```
EDU» subs(f,'x',-0.8165)
ans =
 -0.0887
```

Thus, $f_{max} = 2.0887$ at $x = 0.8165$ and $f_{min} = -0.0887$ at $x = -0.8165$.

Method 2. Using MATLAB's min and max functions.

```
EDU» x = -2:0.05:2;
EDU» f = 1 + 2*x - x.^3;
```

These two lines are identical to those in the script used to plot f above. If you have just run that script you do not need to execute these lines. Now we can find the minimum and maximum values of f by typing *min(f)* and *max(f)*.

```
EDU» min(f)
ans = -3
```

```
EDU» max(f)
ans = 5
```

These are clearly not the answers we obtained above. What went wrong? It is important to understand that *min* and *max* do not find true minima and maxima in the mathematical sense, i.e. they do not find locations where df/dx = 0. Instead, they find the minimum or maximum value of all those calculated in the vector f. To see this, go back and look at the plot of f above and you will see that the minimum and maximum values of f are indeed − 3 and 5. This is another good reason to plot the function first. To avoid the above difficulty we can change the range of x to insure we pick up the true minima and maxima.

```
EDU» x = -1:0.05:1;
```

EDU» f = 1 + 2*x - x.^3;

In most cases, we want not only a minimum or maximum but also the location where they occur. This being the case we should use the following command.

EDU» [f_min,i] = min(f)

f_min = -0.0880

i = 5

The command [f_min, i] = min(f) finds the minimum value of f and assigns it to the variable f_min. The index of this value is assigned to the variable i. Now we can find the location of the minimum by printing the 5th value of the vector x.

EDU» x_min=x(i)

x_min = -0.8000

Now we can repeat the above for the maximum.

EDU» [f_max,i] = max(f)

f_max = 2.0880

i = 37

EDU» x_max=x(i)

x_max = 0.8000

You may have noticed at this point that the answers are somewhat different from those found by the first method above. The reason is, once again, that *min* and *max* do not find true minima and maxima. To obtain more accurate results you can use a closer spacing for the range variable (x). To see this, repeat the above for a spacing of 0.01 instead of 0.05.

Solving several equations simultaneously

EDU» eqn1='x^2+y^2=12'

eqn1 =
x^2+y^2=12

EDU» eqn2='x*y=4'

eqn2 =
x*y=4

In the above we have defined two equations which we will now solve for the two unknowns, x and y.

EDU» [x,y] = solve(eqn1,eqn2)

x =

 5^(1/2)-1
-1-5^(1/2)
 5^(1/2)+1
 1-5^(1/2)

y =

 5^(1/2)+1
 1-5^(1/2)
 5^(1/2)-1
-1-5^(1/2)

Note that four solutions have been found, the first being $x = \sqrt{5} - 1$, $y = \sqrt{5} + 1$.
The following example illustrates a case where there are more unknowns than there are equations. In this case, the specified variables will be solved for in terms of the others.

EDU» f='x^2 + a*y^2 = 0'

f =
x^2 + a*y^2 = 0

EDU» g='x-y = b'

g =
x-y = b

EDU» [x,y] = solve(f,g)

x =

1/2/(a+1)*(-2+2*(-a)^(1/2))*b+b
1/2/(a+1)*(-2-2*(-a)^(1/2))*b+b

y =

1/2/(a+1)*(-2+2*(-a)^(1/2))*b
1/2/(a+1)*(-2-2*(-a)^(1/2))*b

In this case, there are two solutions. Now let's consider an example with three equations.

EDU» eqn1='x^2+y^2=12'

eqn1 =
x^2+y^2=12

EDU» eqn2='x*y=4'

eqn2 =
x*y=4

EDU» eqn3='x-y=z'

eqn3 =
x-y=z

EDU» [x,y,z] = solve(eqn1,eqn2,eqn3)

x =

5^(1/2)+1
 1-5^(1/2)
 5^(1/2)-1
-1-5^(1/2)

y =

 5^(1/2)-1
-1-5^(1/2)
5^(1/2)+1
 1-5^(1/2)

z =

 2
 2
-2
-2

FORCE SYSTEMS

2

This chapter introduces the basic properties of forces and moments in two and three dimensions. Problem 2.1 is a 2D problem that investigates the effects of the orientation of a force upon the resultant of two forces. The problem illustrates how the automatic substitution performed by a computer can simplify what might normally be a rather tedious algebra problem. Problem 2.2 and 2.3 are parametric studies involving 2D moment and equivalent force-couple systems respectively. Problems 2.4 and 2.5 are 3D problems. Problem 2.5 involves an interesting design application. The *solve* command is used to determine the suitable range for a dimension given that the moment must be less than a specified value. Problem 2.6 is another 2D problem that illustrates how to find the maximum value of a moment using three different approaches: (1) geometry, (2) symbolic *diff* and *solve* to find the location of the maximum by setting the derivative equal to zero, and (3) MATLAB's *max* function.

2.1 Problem 2/20 (2D Rectangular Components)

It is desired to remove the spike from the timber by applying force along its horizontal axis. An obstruction A prevents direct access, so that two forces, one 400 lb and the other **P** are applied by cables as shown. Here we want to investigate the effects of the distance between the spike and the obstruction on the two forces so replace 8" by d in the figure. Determine the magnitude of **P** necessary to insure a resultant **T** directed along the spike. Also find T.   Plot P and T as a function of d letting d range between 2 and 12 inches.

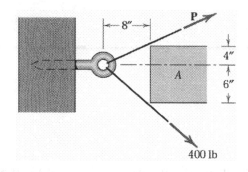

Problem Formulation

The two forces and their resultant are shown on the diagram to the right. The horizontal component of the resultant is T while the vertical component is 0. Thus,

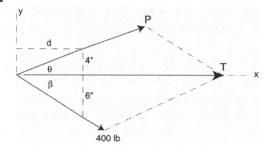

$$T = \Sigma F_x = P\cos\theta + 400\cos\beta$$
$$0 = \Sigma F_y = P\sin\theta - 400\sin\beta$$

These two equations can be easily solved for P and T.

$$P = \frac{400\sin\beta}{\sin\theta} \qquad\qquad T = 400\left(\sin\beta\cot\theta + \cos\beta\right)$$

At this point we have P and T as functions of β and θ. From the figure above we can relate β and θ to d as follows.

$$\theta = \tan^{-1}(4/d) \qquad\qquad \beta = \tan^{-1}(6/d)$$

One nice thing about using a computer is that it will not be necessary to substitute these results into those above to get P and T explicitly as functions of d. The computer carries out this substitution automatically.

MATLAB Script

```
d=2:0.05:12;
theta=atan(4./d); beta=atan(6./d);
P=400*sin(beta)./sin(theta);
T=400*(sin(beta).*cot(theta)+cos(beta));
plot(d,P,d,T)
xlabel('d (in)'); ylabel('Force (lb)');
```

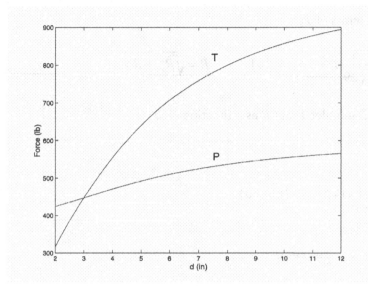

2.2 Problem 2/53 (2D Moment)

The masthead fitting supports the two forces shown. Here we want to investigate the effects of the orientation of the 5 kN load so replace the angle 30° by θ. Find the magnitude of **T** which will cause no bending of the mast (zero moment) at O. For this T, determine the magnitude of the resultant **R** of the two forces. Plot T and R as a function of θ letting q vary between 0 and 180°.

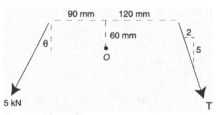

Problem Formulation

Setting the moment about O to zero yields (note that $\sqrt{2^2 + 5^2} = \sqrt{29}$)

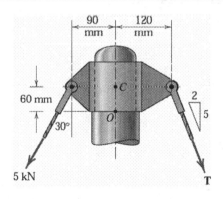

$$M_0 = 5\left[\cos\theta(90) + \sin\theta(60)\right] - T\left[\frac{5}{\sqrt{29}}(120) + \frac{2}{\sqrt{29}}(60)\right] = 0$$

Solving, $T = \dfrac{5\left[\cos\theta(90) + \sin\theta(60)\right]}{\dfrac{5}{\sqrt{29}}(120) + \dfrac{2}{\sqrt{29}}(60)} = \dfrac{5\sqrt{29}}{24}(3\cos\theta + 2\sin\theta)$

Now we can find the resultant R in terms of T,

$$R_x = \frac{2}{\sqrt{29}}T - 5\sin\theta \;\; (\rightarrow) \quad R_y = \frac{5}{\sqrt{29}}T + 5\cos\theta \;\; (\downarrow) \qquad R = \sqrt{R_x^2 + R_y^2}$$

We'll let MATLAB substitute for T and plot T and R as a function of θ.

MATLAB Script

```
theta=0:0.01:pi;
T=5*sqrt(29)/24*(3*cos(theta)+2*sin(theta));
Rx=2/sqrt(29)*T-5*sin(theta);
Ry=5/sqrt(29)*T+5*cos(theta);
R=sqrt(Rx.^2+Ry.^2);
th=theta*180/pi; %converts to degrees
plot(th,T,th,R)
xlabel('theta (degrees)')
ylabel('Force (kN)')
```

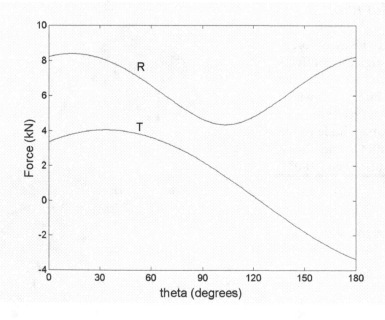

2.3 Problem 2/88 (2D Resultants)

The directions of the two thrust vectors of an experimental aircraft can be independently changed form the conventional forward direction within limits. Here we want to investigate the effects of the orientation of one of the thrusts so change the 15° angle in the figure to θ. For this case determine the equivalent force-couple system at O. Then replace this force-couple system by a single force and specify the point on the x-axis through which the line of action of this resultant passes. Plot (a) the ratio of the magnitude of the resultant to that of the thrust (R/T) and (b) the distance x where the resultant intersects the x-axis as functions of θ for θ between 0 and 30°.

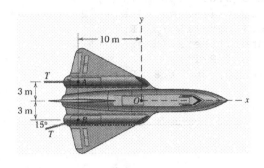

Problem Formulation

For an equivalent force-couple system at O we have,

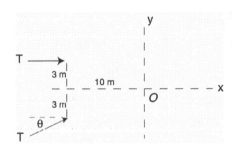

$$R_x = T + T\cos\theta = T(1 + \cos\theta) \quad R_y = T\sin\theta$$

$$R = \sqrt{R_x^2 + R_y^2} = T\sqrt{(1 + \cos\theta)^2 + (\sin\theta)^2}$$

$$\frac{R}{T} = \sqrt{(1 + \cos\theta)^2 + (\sin\theta)^2}$$

$$M_0 = -T(3) + T\cos\theta(3) - T\sin\theta(3) = T(3\cos\theta - 3 - 10\sin\theta) \circlearrowleft$$

Now we want a single resultant force that is equivalent to the force-couple system above. The easiest way to visualize this is to imagine moving the resultant from O to the right along the x-axis. As you do this you generate a counter-clockwise moment about O of $R_y x$ (R_x does not produce a moment about O). The condition for determining x is thus $M_O = R_y x$.

$$x = \frac{M_O}{R_y} = \frac{(3\cos\theta - 3 - 10\sin\theta)}{\sin\theta}$$

MATLAB Scripts

```
%%%%%%%%% Script #1 %%%%%%%%%%%%
% This script plots R/T versus theta
% Let Rp be the ratio R/T
theta=0:0.01:30*pi/180;
Rp=sqrt((1+cos(theta)).^2+sin(theta).^2);
plot(theta*180/pi,Rp)
xlabel('theta (degrees)')
ylabel('R/T')
axis([0, 30, 1.8, 2])
```

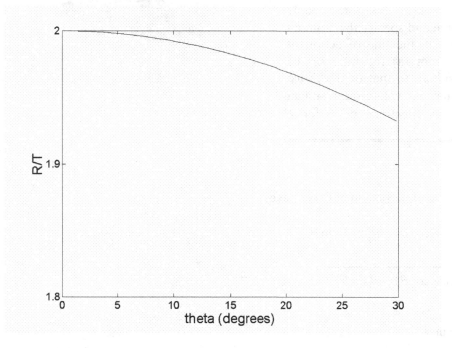

```
%%%%%%%%%% Script #2 %%%%%%%%%%%%%
% This script plots x versus theta
%
theta=0:0.01:30*pi/180;
x=(3*cos(theta)-3-10*sin(theta))./sin(theta);
plot(theta*180/pi,x)
xlabel('theta (degrees)')
ylabel('x (m)')
axis([0, 30, -11, -10])
```

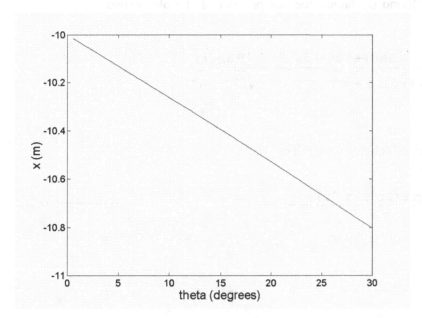

2.4 Problem 2/105 (3D Rect. Components)

The rigid pole and cross-arm assembly is supported by the three cables shown. A turnbuckle at D is tightened until it induces a tension T in CD of 1.2 kN. Let the distance OD be d instead of 3 m and determine the magnitude T_{GF} of the projection of **T** onto line GF. Plot T_{GF} as a function of d letting d vary between 0 and 20 meters.

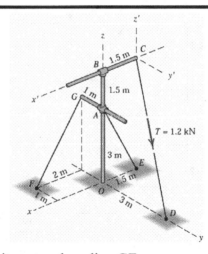

Problem Formulation

First we write **T** as a Cartesian vector and then find a unit vector along line GF.

$$\mathbf{T} = T\frac{1.5\mathbf{i} + d\mathbf{j} - 4.5\mathbf{k}}{\sqrt{(1.5)^2 + d^2 + (4.5)^2}} = (1.2)\frac{1.5\mathbf{i} + d\mathbf{j} - 4.5\mathbf{k}}{\sqrt{22.5 + d^2}}$$

$$\mathbf{n}_{GF} = \frac{2\mathbf{i} - 3\mathbf{k}}{\sqrt{(2)^2 + (3)^2}} = \frac{2\mathbf{i} - 3\mathbf{k}}{\sqrt{13}}$$

The projection T_{GF} is now found by taking the dot product of the above two vectors.

$$T_{GF} = \mathbf{T} \cdot \mathbf{n}_{GF} = \frac{1.2(1.5)(2) + 1.2d(0) + 1.2(-4.5)(-3)}{\sqrt{13}\sqrt{22.5 + d^2}} = \frac{19.8/\sqrt{13}}{\sqrt{22.5 + d^2}}$$

MATLAB Script

```
% This script plots Tgf (the progection of T
% along GF) versus d.
d=0:0.01:20;
Tgf=19.8/sqrt(13)./sqrt(22.5+d.^2);
plot(d,Tgf)
xlabel('d (m)')
ylabel('Tgf (kN)')
```

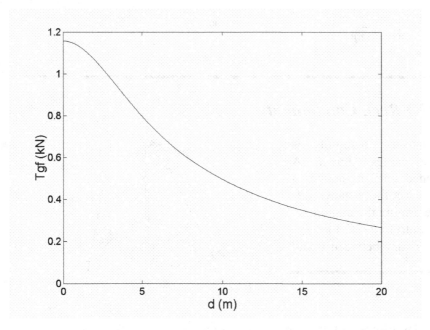

2.5 Sample Problem 2/13 (3D Moment)

A tension **T** of magnitude 10 kN is applied to the cable attached to the top A of the rigid mast and secured to the ground at B. Let the coordinates of point B be $(x_B, 0, z_B)$ instead of $(12, 0, 9)$ as in the figure. Find a general expression for $\mathbf{M_O}$ (the moment of **T** about the base O) as a function of x_B and z_B. (a) Plot the magnitude of $\mathbf{M_O}$ (M_O) and its components about the x and z axes (M_x and M_z) as a function of x_B for $z_B = 9$ m. For this case, determine the suitable range for x_B if M_O is not to exceed 100 kN•m. (b) Find a simple way to explain to a construction worker (who has not taken this course) all safe locations for B if M_O cannot exceed 100 kN•m.

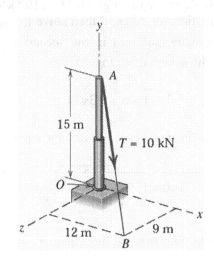

Problem Formulation

First, we express **T** as a Cartesian vector and then find the moment by taking a cross product.

$$\mathbf{T} = T\mathbf{n}_{AB} = 10\left[\frac{x_B\mathbf{i} - 15\mathbf{j} + z_B\mathbf{k}}{\sqrt{x_B^2 + 15^2 + z_B^2}}\right]$$

$$\mathbf{M_O} = \mathbf{r}_{OA} \times \mathbf{T} = 15\mathbf{j} \times 10\left[\frac{x_B\mathbf{i} - 15\mathbf{j} + z_B\mathbf{k}}{\sqrt{x_B^2 + 15^2 + z_B^2}}\right] = 150\left[\frac{z_B\mathbf{i} - x_B\mathbf{k}}{\sqrt{x_B^2 + 15^2 + z_B^2}}\right]$$

(a) M_x and M_z are the scalar components of $\mathbf{M_O}$ in the x and z-directions respectively.

$$M_x = \frac{150z_B}{\sqrt{x_B^2 + 15^2 + z_B^2}} \quad M_z = \frac{-150x_B}{\sqrt{x_B^2 + 15^2 + z_B^2}}$$

$$M_O = \sqrt{M_x^2 + M_z^2} = 150\sqrt{\frac{x_B^2 + z_B^2}{x_B^2 + 15^2 + z_B^2}}$$

After substituting $z_B = 9$, M_O, M_x, and M_z can be plotted as a function of x_B. The result can be found following the first MATLAB script below. To find the acceptable range of x_B given $M_O \leq 100$ kN•m, substitute $M_O = 100$ and $z_B = 9$ into the equation for M_O and then solve for x_B. The symbolic *solve* command is used to obtain the solution in the second script below. The result is $x_B = \pm 3\sqrt{11}$ m. Thus, the acceptable range is

$$-3\sqrt{11} \leq x_B \leq 3\sqrt{11} \ \text{m}$$

(b) Substituting $M_O = 100$ into the equation above gives,

$$100 = 150\sqrt{\frac{x_B^2 + z_B^2}{x_B^2 + 15^2 + z_B^2}} \ .$$

Squaring both sides of this equation and rearranging terms yields

$$x_B^2 + z_B^2 = 180$$

which is a circle of radius $\sqrt{180} = 13.42$ m (44 ft). Thus, the construction worker should be instructed to keep B within 44 feet of the mast.

MATLAB Scripts

```
%%%%%%%%%%%% Script #1 %%%%%%%%%%%%%%%%%
% This script plots Mx, My, and Mo as a
% function of xb for zb = 9 m
xb=-15:0.1:15;
d=inline('sqrt(xb.^2+15^2+9^2)'); % the denominator
Mx=150*9./d(xb);
Mz=-150*xb./d(xb);
Mo=sqrt(Mx.^2+Mz.^2);
plot(xb,Mx,xb,Mz,xb,Mo)
xlabel('xB (m)')
ylabel('M (kN m)')
```

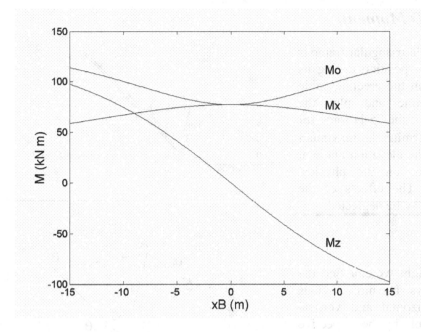

```
%%%%%%%%%%%% Script #2 %%%%%%%%%%%%%%%%%
% This script solves for the value of xb
% for which Mo = 100 kN m when zb = 9 m.
Mo='150*sqrt((xb^2+9^2)/(xb^2+15^2+9^2))=100'
solve(Mo,'xb')
%%%%%%%%%%%%%%%%%%%%%%%%%%%%%%%%%%%%%%%%%%
```

Running this script yields the following,

Mo =
150*sqrt((xb^2+9^2)/(xb^2+15^2+9^2))=100

ans =
 3*11^(1/2)
-3*11^(1/2)

Thus, $M_O = 100$ kN m when $x_B = \pm 3\sqrt{11}$ m.

2.7 Problem 2/181 (2D Moment)

A flagpole with attached light triangular frame is shown here for an arbitrary position during its raising. The 75-N tension in the erecting cable remains constant. Determine and plot the moment about the pivot O of the 75-N force for the range $0 \leq \theta \leq 90°$. Determine the maximum value of this moment and the elevation angle at which it occurs; comment on the physical significance of the latter. The effects of the diameter of the drum at D may be neglected.

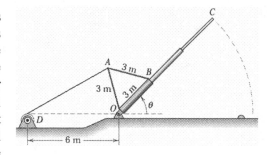

Problem Formulation

From the diagram to the right we can find the moment about O in two ways. The most obvious is to resolve T into horizontal and vertical components and then multiply by the respective moment arms. The result is,

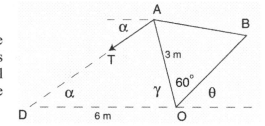

$$M_O = T\cos\alpha(3\sin\gamma) + T\sin\alpha(3\cos\gamma) = 3T(\cos\alpha\sin\gamma + \sin\alpha\cos\gamma)$$

A simpler expression can be found by first sliding T along its line of action to D. Once this has been done we see that only the vertical component ($T\sin\alpha$) produces a moment about O with moment arm of 6 m. Thus,

$$M_O = 6T\sin\alpha$$

Here we will use the second, simpler, expression. As an exercise you may want to verify that the two expressions yield identical results.

Now we have to do a little geometry to relate α to θ. From the diagram we have,

$$\gamma + 60^0 + \theta = 180^0 \qquad \gamma = 120^0 - \theta \qquad \gamma = \frac{2\pi}{3} - \theta \text{ (radians)}$$

$$AD = \sqrt{6^2 + 3^2 - 2(6)(3)\cos\gamma} \qquad\qquad \text{(law of cosines)}$$

$$\frac{\sin\alpha}{3} = \frac{\sin\gamma}{AD} \qquad \alpha = \sin^{-1}\left(\frac{3\sin\gamma}{AD}\right) \qquad \text{(law of sines)}$$

This problem illustrates very well one of the advantages of a computer solution. At this point you should be sure that you understand that we now have the moment as a function of θ. The reason is that we have M_O as a function of α, α as a function of γ and γ as a function of θ. It is not necessary to actually make the substitutions ourselves; the computer will do that for us.

The maximum moment and the elevation at which it occurs are determined in the worksheet below in two ways. The results are M_{Omax}= 225 N-m at $\theta = \pi/3 = 60°$.

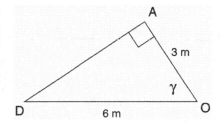

It turns out that we could have anticipated this result with some simple geometry. First, we know that the maximum moment will occur when **T** is perpendicular to *OA*. This yields the right triangle shown to the right. From this diagram we see that $\cos\gamma = 3/6$ so that $\gamma = 60°$. From our results above, $\theta = 60°$ when $\gamma = 60°$. Also, for this orientation, $M_O = 3(75) = 225$ N-m.

MATLAB Scripts

```
%%%%% Script #1 %%%%%
% This script plots Mo (the moment of T
% about O) as a function of theta
theta=0:.01:pi/2;
gamma=2*pi/3-theta;
AD=sqrt(6^2+3^2-2*6*3*cos(gamma));
alpha=asin(3*sin(gamma)./AD);
Mo=6*75*sin(alpha);
plot(theta,Mo)
xlabel('theta (rads)')
ylabel('Mo (N m)')
```

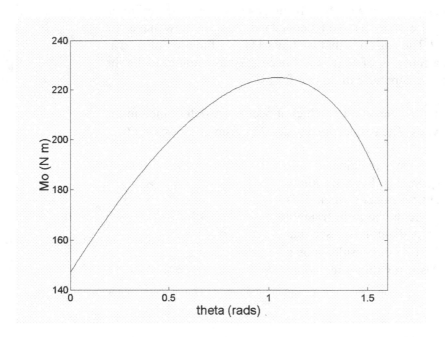

The maximum moment and the elevation at which it occurs will now be determined in two ways.

Method 1. Setting the derivative equal to zero.

Here, the angle θ is determined by solving the equation $dM_O / d\theta = 0$. Substitution of this result into M_O gives the maximum moment. Since the derivative is taken with respect to θ, we first have to find M_O explicitly as a function of θ. This is the most tedious part of the solution. After substituting the expression for γ into that for α and then substituting this result into the expression for M_O we obtain, after some simplification,

$$M_O = \frac{450 \sin\left(\dfrac{\pi}{3} + \theta\right)}{\sqrt{5 + 4\cos\left(\dfrac{\pi}{3} + \theta\right)}}$$

Now we use this in a MATLAB worksheet as follows.

EDU» M = '450*sin(pi/3+theta)/sqrt(5+4*cos(pi/3+theta))'
M =
450*sin(pi/3+theta)/sqrt(5+4*cos(pi/3+theta))

EDU» diff(M,'theta')
 ans =
450*cos(1/3*pi+theta)/(5+4*cos(1/3*pi+theta))^(1/2)+900*sin(1/3*pi+theta)^2/(
5+4*cos(1/3*pi+theta))^(3/2)

The above is the derivative of M_O with respect to θ. Now we set this result equal to zero and name the equation f. This equation is then solved for θ. To reduce the amount of typing, you can copy the above expression into the clipboard and then paste it into the expression for f.

EDU»
f='450*cos(1/3*pi+theta)/(5+4*cos(1/3*pi+theta))^(1/2)+900*sin(1/3*pi+theta)^
2/(5+4*cos(1/3*pi+theta))^(3/2)=0'

f =
450*cos(1/3*pi+theta)/(5+4*cos(1/3*pi+theta))^(1/2)+900*sin(1/3*pi+theta)^2/(
5+4*cos(1/3*pi+theta))^(3/2)=0

EDU» solve(f, 'theta')
ans =
 -1/3*pi+atan(i*3^(1/2),-2)
 -1/3*pi+atan(-i*3^(1/2),-2)
             1/3*pi
              -pi

From the above we see that MATLAB has found four solutions to the equation. The first two are imaginary and can be ignored. Of the two real solutions, only one (pi/3 = 60) is in the range plotted. We'll go ahead and evaluate Mo at both angles. The result indicates that the second real solution (-180 degrees) is a minimum.

EDU» subs(M,'theta',pi/3)
ans = 225.0000

EDU» subs(M,'theta', -pi)
ans = -225.0000

Method 2. Using MATLAB's max function.

To illustrate this approach, first run the MATLAB script above (script #1). Now type max(Mo) at a prompt in the worksheet. You will see the following result.

EDU» max(Mo)

ans =
         224.9991

This is pretty close to the value calculated above. It is not exact since *max* does not calculate a true maximum of a function in the mathematical sense. Instead, it determines the maximum value of all those calculated in the vector Mo. To obtain more accurate results we can modify the script to calculate Mo at a finer spacing of θ. To reduce the size of the array you can zero in on the maximum point by looking at the plot above, which shows the maximum to occur somewhere between 0.9 and 1.1 radians. Here is the modified script.

```
%%%%% Script #2 %%%%%
% This script is used to obtain a more
% accurate estimate of the maximum moment
theta=0.9:.001:1.1;
gamma=2*pi/3-theta;
AD=sqrt(6^2+3^2-2*6*3*cos(gamma));
alpha=asin(3*sin(gamma)./AD);
Mo=6*75*sin(alpha);
```

Now, after running this script we get

EDU» max(Mo)

ans =
         225.0000

Now we have obtained the exact result to within the accuracy of the answer printed. Since we want both the maximum moment and its location it is better to use the following command

EDU» [y, i] = max(Mo)

y =
    225.0000

i =
    148

This command returned the maximum value as y and the index of that value as i. We can now find the angle θ by printing out the 148th term in theta as follows.

EDU» theta(i)

ans =
    1.0470

Or, in degrees,

EDU» theta(i)*180/pi

ans =
    59.9887

This answer shows that the approach yields only approximate results.

EQUILIBRIUM

3

This chapter considers equilibrium of two and three dimensional structures and is the foundation for the study of engineering Statics. Problems 3.1 through 3.3 are two dimensional equilibrium problems while problem 3.4 involves equilibrium in three dimensions. Problem 3.1 and 3.2 are rather simple parametric studies. Problem 3.3 illustrates how to find the maximum and minimum values of a force by using (a) MATLAB's *max* and *min* functions and (b) Maple's symbolic *diff* and *solve* commands. Problem 3.4 studies the effects of the location of a ring support upon several support reactions in a three dimensional structure. The specific location where two of the reaction forces are equal is determined in two ways. The first is a typical graphical approach while the second involves a clever application of MATLAB's *min* function. A method for increasing the accuracy of graphical solutions is also discussed.

3.1 Problem 3/26 (2D Equilibrium)

The indicated location of the center of gravity of the 3600-lb pickup truck is for the unladen condition. A load W_L whose center of gravity is x inches behind the rear axle is added to the truck. Find the relationship between W_L and x if the normal forces under the front and rear wheels are to be equal. For this case, plot W_L as a function of x for x ranging between 0 and 50 inches.

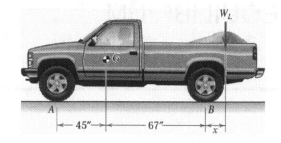

Problem Formulation

The free-body diagram for the truck is shown to the right. Normally, the two normal forces under the wheels would not be identical, of course. Here we want the relationship between the weight W_L and its location (x) which results in these two forces being equal. This relationship is found from the equilibrium equations.

$$\Sigma M_B = 0 = 3600(67) - N(112) - W_L x = 0$$

$$\Sigma F_y = 0 = N + N - 3600 - W_L = 0$$

The second equation can be solved for N and then substituted into the first equation to yield the required relation between W_L and x. This relation can then be solved for W_L with the following result.

$$W_L = \frac{39600}{56 + x}$$

MATLAB Script

```
x=0:0.05:50;
WL=39600./(56+x);
plot(x,WL)
xlabel('x (in)')
ylabel('Load weight (lb)')
```

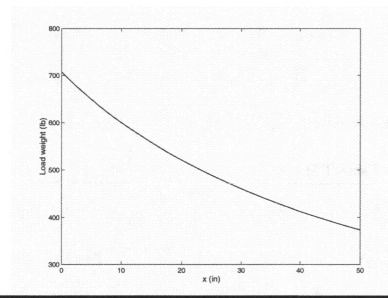

3.2 Problem 3/37 (2D Equilibrium)

The uniform 18-kg bar OA is held in the position shown by the smooth pin at O and the cable AB. Determine the tension T in the cable and the magnitude of the external pin reaction at O in terms of the angle θ. Plot the forces as a function of θ for $15^0 \le \theta \le 90^0$.

Problem Formulation

The free-body diagram for the bar is shown to the right. Before proceeding to the equilibrium equations we must first find the angle α in terms of θ. From the law of cosines and law of sines,

$$AB = \sqrt{1.2^2 + 1.5^2 - 2(1.2)(1.5)\cos(\pi - \theta)}$$

$$\frac{\sin \alpha}{1.2} = \frac{\sin(\pi - \theta)}{AB}$$

Observing that $\cos(\pi - \theta) = -\cos\theta$

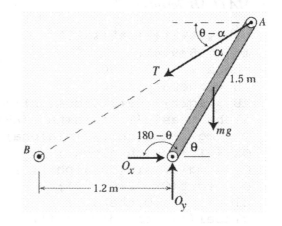

and $\sin(\pi - \theta) = \sin \theta$,

$$AB = \sqrt{3.69 + 3.60 \cos \theta}$$

$$\alpha = \sin^{-1}\left(\frac{1.2 \sin \theta}{AB}\right)$$

The equilibrium equations can now be written,

$$\circlearrowleft \Sigma M_O = 0 = mg\left(\frac{1.5}{2}\cos\theta\right) - T \sin \alpha (1.5)$$

$$\Sigma F_x = 0 = O_x - T \cos(\theta - \alpha)$$

$$\Sigma F_y = 0 = O_y - mg - T \sin(\theta - \alpha)$$

The three equations above can be solved for the forces,

$$T = \frac{mg}{2}\frac{\cos \theta}{\sin \alpha}$$

$$O_x = T \cos(\theta - \alpha) \qquad\qquad O_y = mg + T \sin(\theta - \alpha)$$

$$O = \sqrt{O_x^2 + O_y^2}$$

MATLAB Script

```
%%%%%%%%%%%%%%%%%%%%%%%%%
mg = 18*9.81;
theta = 15:0.1:90;
th = theta*pi/180;
AB = sqrt(3.69+3.6*cos(th));
alpha = asin(1.2*sin(th)./AB);
T = mg/2*cos(th)./sin(alpha);
Ox = T.*cos(th-alpha);
Oy = mg+T.*sin(th-alpha);
O = sqrt(Ox.^2+Oy.^2);
plot(theta,O,theta,T)
xlabel('theta (degrees)')
ylabel('Force (N)')
```

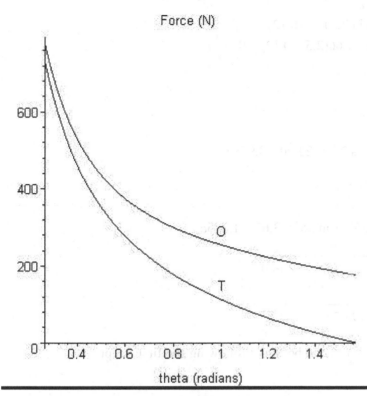

Force (N)

theta (radians)

3.3 Sample Problem 3/4 (2D Equilibrium)

Let the 10 kN load be located a distance c
(meters) to the left of B. Find the magnitude T of
the tension in the supporting cable and the
magnitude of the force on the pin at A in terms
of c. (a) Plot T and A as a function of c letting c
range between 0 and 5 m. (b) Find the minimum
and maximum values for T and A when c varies
between 0 and 5 m. The beam AB is a standard
0.5-m I-beam with a mass of 95 kg per meter of
length.

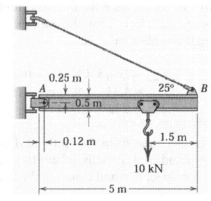

Problem Formulation

The free-body diagram is shown in the figure
to the right. The weight of the beam is
(95)(5)(9.81) = 4660 N or 4.66 kN. The
weight acts at the center of the beam. The
equilibrium equations are now written from
the free-body diagram.

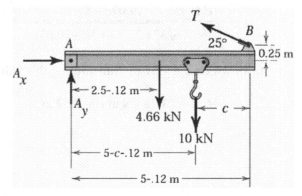

$[\Sigma M_A = 0]$ $(T\cos 25°)0.25 + (T\sin 25°)(5 - 0.12)$
$$- 10(5 - c - 0.12) - 4.66(2.5 - 0.12) = 0$$

from which $T = 26.15 - 4.367c$

$[\Sigma F_x = 0]$ $A_x - T\cos 25° = 0$

$$A_x = (26.15 - 4.367c)\cos 25° = 23.70 - 3.958c$$

$[\Sigma F_y = 0]$ $A_y + T\sin 25° - 4.66 - 10 = 0$

$$A_y = 5.66 - (26.15 - 4.367c)\sin 25° = 3.61 + 1.846c$$

$$A = \sqrt{A_x^2 + A_y^2} = \sqrt{(23.7 - 3.958c)^2 + (3.61 + 1.846c)^2}$$

$$A = \sqrt{574.7 - 174.3c + 19.07c^2}$$

(a) The plot of T and A as a function of c can be found following the first script below.

(b) The locations c for the maximum values of T and A and the minimum value for T are clear from the plot below. Substituting these values of c into the equations above yields,

$T_{max} = 26.15$ kN at $c = 0$
$T_{min} = 4.31$ kN at $c = 5$ m
$A_{max} = 23.97$ kN at $c = 0$

It is also clear from the plot that A goes through a minimum somewhere between $c = 4$ and 5 m. Exactly where this occurs can be determined by differentiating A with respect to c and equating the result to zero.

$$\frac{dA}{dc} = \frac{19.07c - 87.15}{\sqrt{574.7 - 174.3c + 19.07c^2}} = 0$$

from which $c = 4.57$ m. Substituting this value into A yields,

$A_{min} = 13.28$ kN at $c = 4.57$ m

As we will see in the results below, MATLAB can be used to obtain many of the detailed aspects of the solution outlined above. We will also illustrate an alternative method for determining A_{min} and its location.

MATLAB Script and Worksheet

```
%%%%%%%%%%%%%%%%%%%%%%%%%%%%%%%%%%%%%%%%%%
% This script plots T and A as a function
% of c.
c=0:0.01:5;
T=26.15-4.367*c;
A=sqrt(574.7-174.3*c+19.07*c.^2);
plot(c,T,c,A)
xlabel('c (m)')
ylabel('Force (kN)')
```

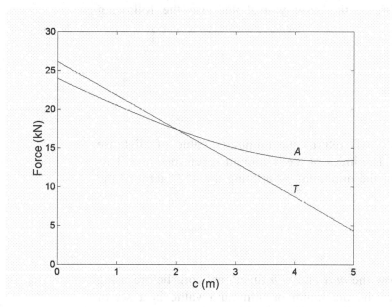

We will use two methods for finding the maximum and minimum values for T and A and their locations.

Method 1. Using MATLAB's max and min functions.

The locations for the maximum values of T and A and the minimum value for T are clear from the plot. We can obtain the associated maximum and minimum values by substituting these locations (values for c) into the equations above. Another approach (which will also find the minimum value for A) is to use MATLAB's *max* and *min* functions. First run the script above and then type the following at a prompt in a worksheet.

EDU» Tmax=max(T)
Tmax = 26.1500

EDU» Tmin=min(T)
Tmin = 4.3150

EDU» Amax=max(A)
Amax = 23.9729

EDU» Amin=min(A)
Amin = 13.2825

Once again, the locations for the maximum values of T and A and the minimum value for T are clear from the plot (they occur at either c = 0 or c = 5 m). The location for Amin is not clear. In this case you should use the following command.

EDU» [y, i] = min(A)
y = 13.2825

i = 458

Following this command, MATLAB returns the minimum value of all those calculated in the vector A as y. The index of the minimum is returned as i. This allows us to find the location of the minimum by printing the 458th term of the vector c as follows.

EDU» c(i)
ans = 4.5700

It is important to understand how the *max* and *min* functions work before using them. The functions find either the maximum or minimum value in a set of calculated values (the vectors A and T in our case). This value may or may not occur at a relative max or min in the usual, mathematical sense. In other words, MATLAB does not find locations where the slope (derivative) equals zero. This is well illustrated by using the function to find the minimum and maximum values of T as we did above. Referring back to the graph you will see that there are no locations over the range plotted where dT/dc = 0. This brings us to method 2 which does look for maximums or minimums by first determining where the slope is zero.

Method 2. Using symbolic diff *and* solve *to solve the equation dA/dc = 0.*

This method is appropriate only for determining Amin since the other maximums and minimums do not occur at locations where the slope is zero. These can be determined by inspection from the graph as above.

To find the location of Amin we first write A as a Maple expression and then differentiate A with respect to c and set the result equal to zero. The value of c thus determined is then substituted back into A to give Amin.

EDU» A='sqrt(574.7-174.3*c+19.07*c^2)'
A =
sqrt(574.7-174.3*c+19.07*c^2)

EDU» dAdc=diff(A,'c')
 dAdc =
 1/2/(574.7-174.3*c+19.07*c^2)^(1/2)*(-174.3+38.14*c)

EDU» cm=solve(dAdc,'c')
 cm =
 4.5700052438384897745149449396959

Note that when using *solve*, MATLAB will assume that the expression is set equal to zero unless otherwise indicated. This is convenient in the present case since we want the solution for the case dAdc = 0.

EDU» subs(A,'c',cm)
 ans =
 13.282471268535671714984943881577

Thus, A_{min} = 13.28 kN at c = 4.57 m

3.4 Sample Problem 3/7 (3D Equilibrium)

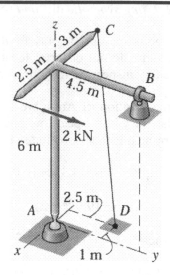

The welded tubular frame is secured to the horizontal x-y plane by a ball and socket joint at A and receives support from the loose-fitting ring at B. Under the action of the 2-kN load, the cable CD prevents rotation about a line from A to B, and the frame is stable in the position shown. Let the distance from the z-axis to the ring at B be c meters so that the coordinates of B are (0, c, 6). Find expressions for the tension T in the cable and the magnitudes of the forces at A and B in terms of c. Plot T, A, and B as a function of c for $0 \leq c \leq 10$ m. Explain why A and B go to infinity as c approaches 0. Find the value of c for which A = B. You may neglect the weight of the frame.

Problem Formulation

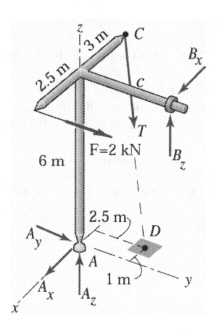

In the solution to the sample problem in your text, moments were summed about the AB axis in order to obtain one equation with one unknown, the tension T. Here we will take a more straightforward approach, summing moments about A and then summing forces.

Summing moments about A in the free-body diagram shown to the right,

$$[\Sigma \mathbf{M}_A = 0] \qquad \mathbf{r}_{AD} \times \mathbf{T} + \mathbf{r}_{AE} \times \mathbf{F} + \mathbf{r}_{AB} \times \mathbf{B} = 0$$

where $\mathbf{T} = \dfrac{T(2\mathbf{i} + 2.5\mathbf{j} - 6\mathbf{k})}{\sqrt{(2)^2 + (2.5)^2 + (-6)^2}} = \dfrac{2T}{\sqrt{185}}(2\mathbf{i} + 2.5\mathbf{j} - 6\mathbf{k})$

$\mathbf{F} = 2\mathbf{j}$ (kN) $\mathbf{B} = B_x \mathbf{i} + B_Z \mathbf{k}$

$\mathbf{r}_{AD} = -\mathbf{i} + 2.5\mathbf{j} \quad \mathbf{r}_{AE} = 2.5\mathbf{i} + 6\mathbf{k} \quad \mathbf{r}_{AB} = c\mathbf{j} + 6\mathbf{k}$

Completion of the vector operations above yields a vector equation whose x, y, and z components give three scalar equations for the summation of moments about the x, y, and z axes respectively,

$$\Sigma M_x = -\frac{6\sqrt{185}}{37}T - 12 + cB_z = 0 \qquad \Sigma M_y = -\frac{12\sqrt{185}}{185}T + 6B_x = 0$$

$$\Sigma M_z = -\frac{3\sqrt{185}}{37}T + 5 - cB_x = 0$$

These three equations can be solved simultaneously to give

$$T = \frac{5\sqrt{185}}{15 + 2c} \qquad B_x = \frac{10}{15 + 2c} \qquad B_z = \frac{6(55 + 4c)}{c(15 + 2c)}$$

The remaining unknowns can be found by summing forces,

$$\Sigma F_x = T_x + B_x + A_x = 0; \; \Sigma F_y = T_y + 2 + A_y = 0;$$

$$\Sigma F_z = T_z + B_z + A_z = 0$$

Substituting the components of **T** (T_x, T_y, T_z) and **B** $(B_x \; B_z)$ into the above yields,

$$A_x = -\frac{30}{15 + 2c} \quad A_y = -\frac{55 + 4c}{15 + 2c} \quad A_z = \frac{6(6c - 55)}{c(15 + 2c)}$$

Finally, $\quad A = \sqrt{A_x^2 + A_y^2 + A_z^2} \quad$ and $B = \sqrt{B_x^2 + B_z^2}$

MATLAB Scripts

```
%%%%%%%%%% Script #1 %%%%%%%%%%%%%%
% This script plots T, A, and B
% as a function of c
c=0:0.02:10;
T=5*sqrt(185)./(15+2*c);
Bx=10./(15+2*c);
Bz=6*(55+4*c)./c./(15+2*c);
Ax=-30./(15+2*c);
Ay=-(55+4*c)./(15+2*c);
Az=6*(6*c-55)./c./(15+2*c);
B=sqrt(Bx.^2+Bz.^2);
A=sqrt(Ax.^2+Ay.^2+Az.^2);
plot(c,T,c,A,c,B)
xlabel('c (m)')
ylabel('Force (kN)')
axis([0,10,0,10])
%%%%%%%%%% end of script %%%%%%%%%
```

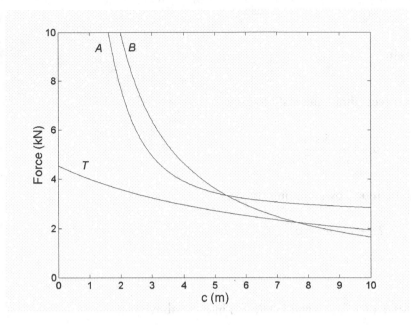

When you run the script you will get an error message for division be zero. The reason for this is that c begins at zero and also appears in the denominator of the z components of A and B. This is the numerical reason that A and B approach infinity as c approaches zero. The physical explanation is that the ring can no longer support the frame when it is located exactly on the z-axis. Since A and B approach infinity as c goes to zero we limit the vertical scale to be between 0 and 10. This is accomplished with the *axis* command in the script above.

Now we illustrate two methods for finding the value of c for which $A = B$.

Method 1. Graphical approach

This is probably the easiest method once you have generated the plot above. All you have to do is find the intersection of the two curves and then drop a vertical line to read the value of c. This is illustrated in the figure below.

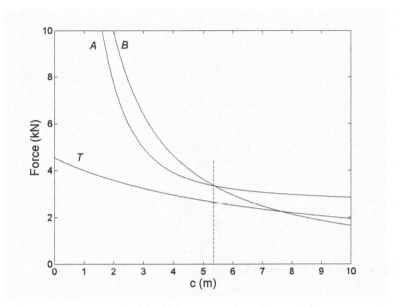

From the above plot we can estimate the value of c to be about 5.3 m. The obvious disadvantage of this method is that it is not very accurate. We can improve the accuracy considerably by limiting the range on c to the neighborhood where $A = B$. The following script illustrates this.

```
%%%%%%%%%% Script #2 %%%%%%%%%%%%%%%
% This script plots A and B as a function
% of c in the neighborhood of the location where
% A = B.
c=5.3:0.001:5.4;
Bx=10./(15+2*c);
Bz=6*(55+4*c)./c./(15+2*c);
Ax=-30./(15+2*c);
Ay=-(55+4*c)./(15+2*c);
Az=6*(6*c-55)./c./(15+2*c);
B=sqrt(Bx.^2+Bz.^2);
A=sqrt(Ax.^2+Ay.^2+Az.^2);
plot(c,A,c,B)
xlabel('c (m)')
ylabel('Force (kN)')
%%%%%%%%%% end of script %%%%%%%%%
```

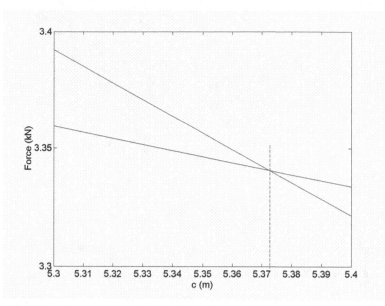

From this plot we obtain a value for c of about 5.373 m, considerably more accurate than that found above.

Method 2. Using MATLAB's min *function*

At first it probably seems odd to try to find the location where $A = B$ using the *min* function. To provide the motivation for the approach consider the following plot of $A - B$ and the absolute value of $A - B$.

```
%%%%%%%%%%% Script #3 %%%%%%%%%%%%%%%
% This script plots A - B and the absolute
% value of A - B as a function of c.
c=0:0.01:10;
Bx=10./(15+2*c);
Bz=6*(55+4*c)./c./(15+2*c);
Ax=-30./(15+2*c);
Ay=-(55+4*c)./(15+2*c);
Az=6*(6*c-55)./c./(15+2*c);
B=sqrt(Bx.^2+Bz.^2);
A=sqrt(Ax.^2+Ay.^2+Az.^2);
C=A-B;
D=abs(C);
plot(c,C,c,D)
xlabel('c (m)')
ylabel('A-B; abs(A-B)')
%%%%%%%%%%% end of script %%%%%%%%%
```

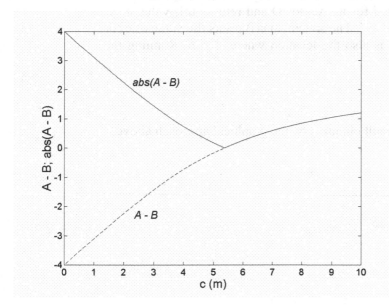

Note that while $A - B$ has no minimum over the range plotted, the absolute value of $A - B$ does have a minimum (equal to zero) at the exact location where $A = B$. Of course, we could drop a vertical from this location to determine c but this would be identical to the graphical approach above.

Instead we will write one more script (most of which is copied and pasted from script # 2 above) which finds the location where $A = B$ using the idea illustrated in the graph above.

```
%%%%%%%%%% Script #4 %%%%%%%%%%%%%%
% This script uses min to find the value
% of c where A = B
c=5.3:0.001:5.4;
Bx=10./(15+2*c);
Bz=6*(55+4*c)./c./(15+2*c);
Ax=-30./(15+2*c);
Ay=-(55+4*c)./(15+2*c);
Az=6*(6*c-55)./c./(15+2*c);
B=sqrt(Bx.^2+Bz.^2);
A=sqrt(Ax.^2+Ay.^2+Az.^2);
D=abs(A - B);
[y, i] = min(D);
c_min = c(i)
%%%%%%%%%% end of script %%%%%%%%%
```

In the above script, D = abs(A - B). The command [y, i] = min(D) determines the minimum value of those calculated for the vector D and returns this value as y. The index of this value is returned as i. The statement c(i) prints the value of c for which D was a minimum which is also the location where $A = B$. Running the script gives the following result,

c_min = 5.3730

Note that this is identical to the result obtained by the graphical approach above.

STRUCTURES

4

This chapter concerns the determination of internal forces in a structure. A structure is an assembly of connected members designed to support or transfer forces. Thus, the internal forces that are generally of interest are the forces of action and reaction between the connected members of the structure. The types of structures considered here are generally classified as trusses, frames or machines.

Problems 4.1 and 4.2 use the methods of joints and sections respectively to analyze a two dimensional truss. Problem 4.1 uses the symbolic solve command to solve two equations for two unknowns. Problem 4.3 studies the effects of the height of a space truss upon the internal reaction forces in three of its members. The method of joints is used to formulate this problem. Problems 4.4 and 4.5 cover frames and machines.

4.1 Problem 4/14 (Trusses. Method of Joints)

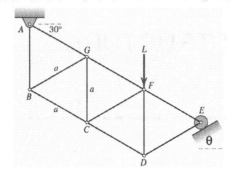

The truss is composed of equilateral triangles of sides a and is loaded and supported as shown. Determine the reaction force on the roller at E and the forces in members FE, and DE as a function of θ. Plot the non-dimensional loads E/L, FE/L, and DE/L for θ between 0 and 45 degrees.

Problem Formulation

First we determine the reaction force at E from a free-body diagram for the entire truss (shown to the right).

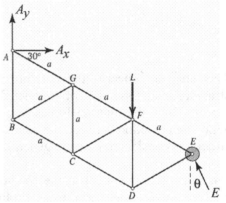

$$\Sigma M_A = -L(2a\cos(30)) + E\cos\theta(3a\cos(30))$$
$$- E\sin\theta(3a\sin(30)) = 0$$

$$E = \frac{2}{3}\frac{L\cos 30}{\cos\theta\cos 30 - \sin\theta\sin 30} = \frac{2}{3}\frac{L\sqrt{3}}{\sqrt{3}\cos\theta - \sin\theta}$$

Note that the non-dimensional force E/L can be easily found by dividing both sides of the equation by L.

$$\frac{E}{L} = \frac{2}{3}\frac{\sqrt{3}}{\sqrt{3}\cos\theta - \sin\theta}$$

Note also that this same equation could be obtained just as easily by setting $L = 1$. Thus one interpretation of a non-dimensional load such as E/L is the force per unit load L.

To obtain the required forces FE and DE we now consider a free-body diagram for joint E. Note that each member is assumed to be in tension. Thus, positive answers will imply tension and negative answers compression.

Joint E

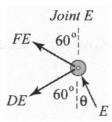

Joint E

$$\Sigma F_x = 0 = -FE\sin 60 - DE\sin 60 - E\sin\theta$$

$$\Sigma F_y = 0 = FE\cos 60 - DE\cos 60 + E\cos\theta$$

Note that after E is substituted that the above are two equations in two unknowns for FE and DE. We will let MATLAB carry out the solution to these equations in the script below. The result is

$$\frac{FE}{L} = -\frac{2}{3}\frac{\sqrt{3}\cos\theta + \sin\theta}{\sqrt{3}\cos\theta - \sin\theta} \qquad\qquad \frac{DE}{L} = \frac{2}{3}$$

Of course, the fact that the force in member DE is $2L/3$ independent of θ is probably surprising.

MATLAB Script #1

```
%%%%%%%%%%%%%%%% Script #1 %%%%%%%%%%%%%%%%%%%%%
% This script solves the two equilibrium equations
% obtained by summing forces for joint E. Note that
% x = FE, y = DE, and z = E. Also notice how the
% result for E is substituted into equations 1 and 2
syms x y z E
E='2/3*sqrt(3)/(sqrt(3)*cos(theta)-sin(theta))';
eqn1='x*sin(60*pi/180)+y*sin(60*pi/180)+z*sin(theta)=0';
eqn2='x*cos(60*pi/180)-y*cos(60*pi/180)+z*cos(theta)=0';
eqn3 = subs(eqn1,z,E);
eqn4 = subs(eqn2,z,E);
[x,y] = solve(eqn3,eqn4)
%%%%%%%%% end of script %%%%%%%%%%%%%%%%%%%%%%%%%
```

Output from script #1

x =
  -2/3*(3^(1/2)+tan(theta))/(3^(1/2)-tan(theta))

y =
  2/3


```
%%%%%%%%%%%%%%% Script #2 %%%%%%%%%%%%%%%%%%%
% This script plots the non-dimensional forces
% at the roller E and in members FE and DE.
th=0:0.01:pi/4;
E = 2/3*sqrt(3)./(sqrt(3)*cos(th)-sin(th));
DE = 2/3;
FE = -2/3*(tan(th)+sqrt(3))./(sqrt(3)-tan(th));
plot(th,E,th,DE,th,FE)
xlabel('theta (rads)')
ylabel('Non-dimensional Force')
%%%%%%%%% end of script %%%%%%%%%%%%%%%%%%%%%%
```

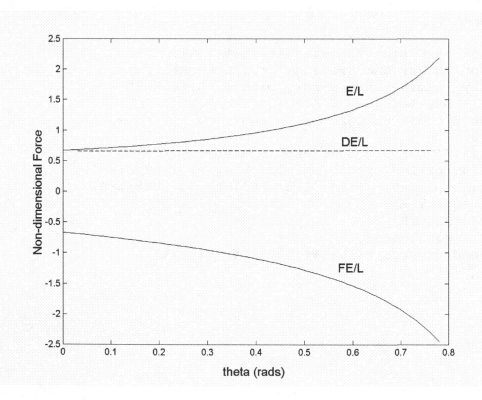

4.2 Problem 4/48 (Trusses. Method of Sections)

In this problem we would like to investigate the effects of the geometry of the upper part of the arched roof truss on some of the forces. We are given a constraint that the vertical distance between the horizontal section BF and the top most point must be 7 m, however we can vary the vertical location of points E and C. Determine the forces in members DE, EI, EF, FI, and HI of the truss. First, write the results in terms of the lengths a and b, then plot the forces as a function of a with the constraint that $a + b$ is always 7 m. Let a vary between 1 and 6 meters.

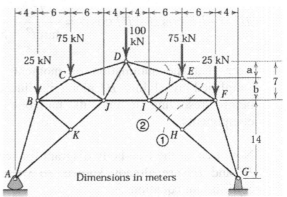

Problem Formulation

First we determine the reaction force at G from a free-body diagram for the entire truss (shown to the right). Due to symmetry,

$$A_y = G_y = (25 + 75 + 100 + 75 + 25)/2 = 150 \text{ kN}$$

From the free-body diagram we also find the angles,

$$\theta = \tan^{-1}\left(\frac{a}{6}\right) \quad \beta = \tan^{-1}\left(\frac{b}{6}\right) \quad \alpha = \tan^{-1}\left(\frac{14}{16}\right)$$

From the free-body diagram for section 1

$$\circlearrowleft \Sigma M_I = 0 = 150(16) - 25(12) + EF \sin \beta (12)$$

$$EF = -\frac{175}{\sin \beta}$$

$$\circlearrowleft \Sigma M_F = 0 = 150(4) - IH \sin \alpha (12)$$

$$IH = \frac{50}{\sin \alpha}$$

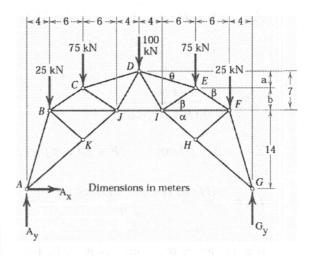

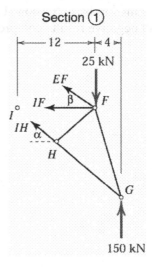

Section ①

$$\Sigma F_x = 0 = -IF - EF\cos\beta - IH\cos\alpha$$

$$IF = -EF\cos\beta - IH\cos\alpha = \frac{175}{\tan\beta} - \frac{50}{\tan\alpha}$$

Note that the summation of moments about F was simplified by sliding IH along its line of action to I before resolving it into its horizontal and vertical components.

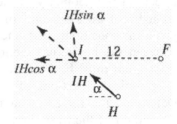

Now look at the free-body diagram for section 2. Forces IF and IH are already known, so we need only two equilibrium equations.

$$\Sigma M_I = 0 = DE\cos\theta(b) + DE\sin\theta(6)$$
$$- 75(6) - 25(12) + 150(16)$$

$$DE = \frac{-1650}{6\sin\theta + b\cos\theta}$$

$$\Sigma F_x = 0 = IH\cos\alpha + IF + EI\cos\beta + DE\cos\theta$$

$$EI = -\frac{IH\cos\alpha + IF + DE\cos\theta}{\cos\beta}$$

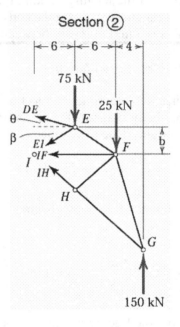

We will allow the computer to make the substitutions of IH, IF, and DE in the last equation. Note that all the members were assumed to be in tension. Thus positive results will imply tension and negative results compression.

MATLAB Script

```
a = 1:0.01:6;
b = 7-a;
alpha = atan(14/16);
beta = atan(b/6);
theta = atan(a/6);
EF = -175./sin(beta);
IH = 50./sin(alpha);
IF = 175./tan(beta)-50./tan(alpha);
DE = -1650./(6*sin(theta)+b.*cos(theta));
EI = -(IH.*cos(alpha)+IF+DE.*cos(theta))./cos(beta);
plot(a,EF,a,IH,a,IF,a,DE,a,EI)
xlabel('a (m)')
ylabel('Force (kN)')
```

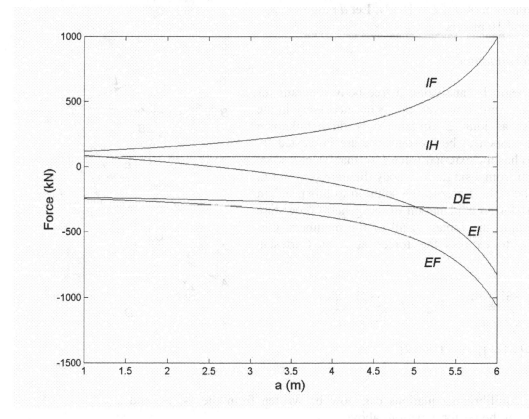

4.3 Sample Problem 4/5 (Space Trusses)

The space truss consists of the rigid tetrahedron *ABCD* anchored by a ball-and-socket connection at *A* and prevented from any rotation about the *x*-, *y*-, or *z*-axes by the respective links 1, 2, and 3. The load *L* is applied at joint *E*, which is rigidly fixed to the tetrahedron by the three additional links. Here we would like to investigate how the forces in a few of the members depend upon the height of the structure so let the height be *d* instead of 4 meters. Solve for the forces in the members at joint *E* and plot these as a function of *d* if $L = 10$ kN. Let *d* range between 2 and 10 meters.

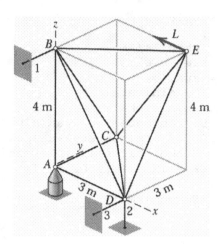

Problem Formulation

We could begin by analyzing a free-body diagram for the entire structure; however, this is not necessary in the present case as joint *E* contains only three unknown forces that happen to be the ones we are interested in. Joint *E* also has the external force *L* so that we can solve this problem from a single free-body diagram for joint *E*, shown to the right. Refer to the solution to this problem in your text for an explanation of the general procedure for determining the forces in all of the members. Our first step is to express the forces at *E* as Cartesian vectors.

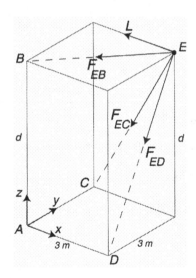

$$\mathbf{F_{EC}} = F_{EC}\frac{-3\mathbf{i} - d\mathbf{k}}{\sqrt{9 + d^2}} \qquad \mathbf{F_{ED}} = F_{ED}\frac{-3\mathbf{j} - d\mathbf{k}}{\sqrt{9 + d^2}}$$

$$\mathbf{F_{EB}} = \frac{F_{EB}}{\sqrt{2}}(-\mathbf{i} - \mathbf{j}) \qquad \mathbf{L} = -L\mathbf{i}$$

The scalar equilibrium equations can now be written from the *x*-, *y*-, and *z*-components of the vector equations above.

$$\Sigma F_x = 0 = -L - \frac{F_{EB}}{\sqrt{2}} - \frac{3F_{EC}}{\sqrt{9 + d^2}} \qquad\qquad \Sigma F_y = 0 = -\frac{F_{EB}}{\sqrt{2}} - \frac{3F_{ED}}{\sqrt{9 + d^2}}$$

$$\Sigma F_z = 0 = -\frac{d}{\sqrt{9+d^2}}\left(F_{EC} + F_{ED}\right)$$

These equations can be solved for the three forces,

$$F_{ED} = -F_{EC} = \frac{L}{6}\sqrt{9+d^2} \qquad F_{EB} - \frac{L}{\sqrt{2}}$$

Referring back to the free-body diagram we see that all three members were assumed to be in tension. Thus, the results indicate that *ED* is in tension while *EC* and *EB* are in compression.

MATLAB Script

```
d=2:0.05:10;
L=10;
Fed=L/6*sqrt(9+d.^2);
Fec=-Fed;
Feb=-L/sqrt(2);
plot(d,Fed,d,Fec,d,Feb)
xlabel('d (m)')
ylabel('Force (kN)')
```

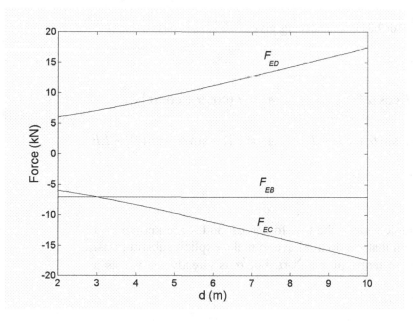

4.4 Problem 4/83 (Frames and Machines)

Determine the magnitude of the pin reaction at A and the magnitude of the force reaction at the rollers as a function of x. The pulleys at C and D are small. Plot the two forces for $0 \le x \le 0.8$ m. What is the direction of the force at the rollers? Does the direction change for the range of x considered?

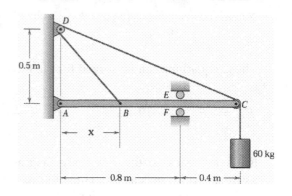

Problem Formulation

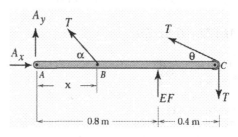

$$\alpha = \tan^{-1}(0.5/x) \qquad \theta = \tan^{-1}(0.5/1.2) = 22.62°$$

$$\circlearrowleft M_A = 0 = Tx\sin\alpha + T(1.2)\sin\theta - T(1.2) + EF(0.8)$$

Note in the Free Body Diagram that we have assumed that force EF acts up. This is equivalent to assuming that the boom is in contact with and pushing down on roller F. Roller F in turn pushes back up on the boom. This assumption is verified by calculating only positive values for the force.

Substituting $\theta = 22.62°$, T = 60(9.81), and solving gives,

$$EF = 543.32 - 735.75x\sin\alpha$$

$$\Sigma F_x = 0 = A_x - T\cos\theta - T\cos\alpha \qquad\qquad A_x = T(\cos\theta + \cos\alpha)$$

$$\Sigma F_y = 0 = A_y + T\sin\alpha + T\sin\theta + EF - T \qquad A_y = T(1 - \sin\alpha - \sin\theta) - EF$$

$$A = \sqrt{A_x^2 + A_y^2}$$

The formulation is now complete since the tow forces EF and A are known as a function of x. Note once again that we do not have to make explicit substitutions. For example, EF is written in terms of x and α, but α is already known as a function of x.

MATLAB Script

```
x = 0.01:0.01:0.8;
alpha = atan(.5./x);
theta = atan(.5/1.2);
T = 60*9.81;
EF = 543.32 - 735.75*x.*sin(alpha);
Ax = T*(cos(theta)+cos(alpha));
Ay = T*(1-sin(alpha)-sin(theta))-EF;
A = sqrt(Ax.^2+Ay.^2);

plot(x,A,x,EF)
xlabel('x (m)')
ylabel('Force (N)')
```

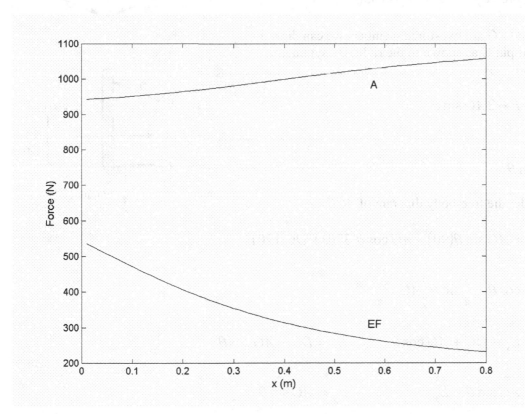

4.5 Problem 4/108 (Frames and Machines)

A lifting device for transporting 135-kg steel drums is shown. Develop expressions for the forces at A, C, and E in terms of L, the length of member AG. Plot these forces as a function L between 345 and 380 mm.

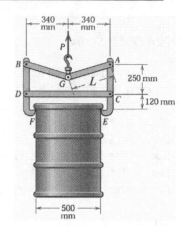

Problem Formulation

$$\theta = \cos^{-1}(340/L) \qquad P = W = 135(9.81) \text{ N}$$

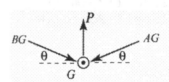

Since BG and AG are two-force members we can draw a free-body of pin G as shown to the right. By symmetry, $AG = BG$, so

$$\Sigma F_y = 0 = P - 2AG\sin\theta$$

$$AG = \frac{P}{2\sin\theta}$$

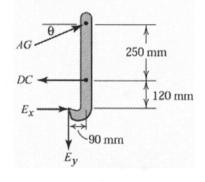

Now consider the free-body diagram of ACE.

$$\circlearrowleft M_E = 0 = AG\sin\theta(90) - AG\cos\theta(370) + DC(120)$$

$$DC = \left(\frac{37}{12}\cos\theta - \frac{3}{4}\sin\theta\right)AG$$

$$\Sigma F_x = 0 = E_x - DC + AG\cos\theta \qquad E_x = DC - AG\cos\theta$$

$$\Sigma F_y = 0 = AG\sin\theta - E_y \qquad E_y = AG\sin\theta = \frac{P}{2}$$

$$E = \sqrt{E_x^2 + E_y^2}$$

The formulation is now complete since all forces are known in terms of θ which is in turn known as a function of *L*.

MATLAB Script

```
L = 345:0.1:380;
theta = acos(340./L); P = 135*9.81;
AG = P/2./sin(theta);
DC = (37/12*cos(theta)-3/4*sin(theta)).*AG;
Ex = DC - AG.*cos(theta);
Ey = P/2;
E = sqrt(Ex.^2+Ey.^2);
plot(L,AG/1000,L,DC/1000,L,E/1000)
xlabel('L (mm)')
ylabel('Force (kN)')
```

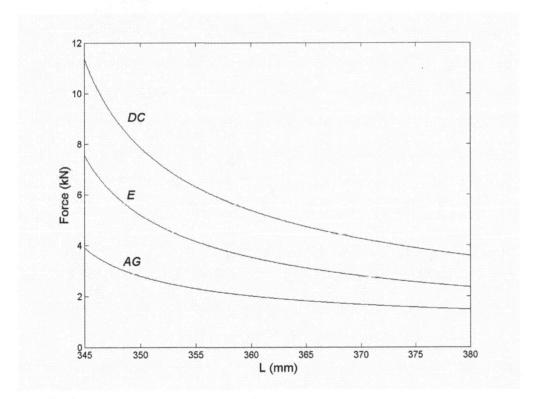

DISTRIBUTED FORCES

5

This chapter considers forces that are distributed over lines, areas, or volumes. Also considered are a few preliminary topics directly related to distributed forces such as center of mass and centroids. The major applications are beams, cables and fluid statics.

Problem 5.1 contains the first example in this booklet of a parametric plot. Problem 5.2 formulates the equilibrium equations for a beam in terms of an integral of the distributed load on the beam. Problem 5.3 obtains the internal shear force V and bending moment M as integrals of the distributed loading function. The symbolic *int* command is used to evaluate the integrals. In problem 5.4, *fzero* is used to obtain a numerical solution for the minimum tension and total length of a flexible cable.

5.1 Sample Problem 5/8 (Composite Bodies)

Let the length of the 40-mm diameter shaft in the bracket-and-shaft assembly be L (mm). The vertical face of the assembly is made from sheet metal that has a mass of 25 kg/m^2. The material of the horizontal base has a mass of 40 kg/m^2, and the steel shaft has a density of 7.83 Mg/m^3. (a) Find expressions for the location of the center of mass ($\bar{Y}$, $\bar{Z}$) in terms of L. (b) Plot $\bar{Y}$ and $\bar{Z}$ as a function of L for $0 \le L \le 200$ mm. (c) Plot $\bar{Z}$ versus $\bar{Y}$ for $0 \le L \le 200$ mm.

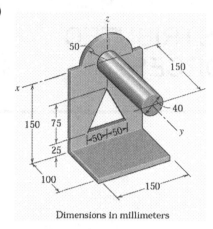

Dimensions in millimeters

Problem Formulation

(a) Refer to the solution to this sample problem in your text and be sure that you understand how the results in the table were obtained. For the problem considered here, the table will remain unchanged except for the last row. For the steel shaft (part 5) we have,

$$m = 0.00984L; \quad \bar{y} = L/2; \quad \bar{z} = 0; \quad m\bar{y} = 0.00492L^2; \quad m\bar{z} = 0.$$

The totals become

$$\Sigma m = 1.166 + 0.00984L; \quad \Sigma m\bar{y} = 30 + 0.00492L^2; \quad \Sigma m\bar{z} = -120.73$$

Thus,

$$\bar{Y} = \frac{\Sigma m\bar{y}}{\Sigma m} = \frac{30 + 0.00492L^2}{1.166 + 0.00984L}; \quad \bar{Z} = \frac{\Sigma m\bar{z}}{\Sigma m} = \frac{-120.73}{1.166 + 0.00984L}$$

(b) The plot of $\bar{Y}$ and $\bar{Z}$ versus L can be found following the first MATLAB script below.

(c) At first sight, plotting $\bar{Z}$ versus $\bar{Y}$ may seem problematic since $\bar{Z}$ is not known explicitly as a function of $\bar{Y}$. Perhaps the most straightforward way of dealing with this apparent difficulty is to solve the first equation above for L in terms of $\bar{Y}$ and then substitute this result into the second equation. Though this approach might seem simple at first, it turns out that there are two solutions, both of which are very messy. There may also be situations where this substitution approach may be impossible, even with symbolic algebra.

Fortunately, most software packages allow plotting two variables without performing a substitution, provided the two variables are expressed in terms of a common parameter (in our case L). Plots of this type are usually called parametric plots. The parametric plot of $\bar{Z}$ versus $\bar{Y}$ can be found following the second MATLAB script below.

MATLAB Scripts

```
%%%%%%%%%%%%%%% Script #1 %%%%%%%%%%%%%
% This script plots the y and z coordinates
% of the center of mass as a function of L
L=0:0.5:200;
sum_mass=1.166+0.00984*L;
yb=(30+0.00492*L.^2)./sum_mass;
zb=-120.73./sum_mass;
plot(L,yb,L,zb)
xlabel('L (mm)')
ylabel('(mm)')
title('plot for part (b)')
%%%%%%%%%%%%%% end of script %%%%%%%%%%
```

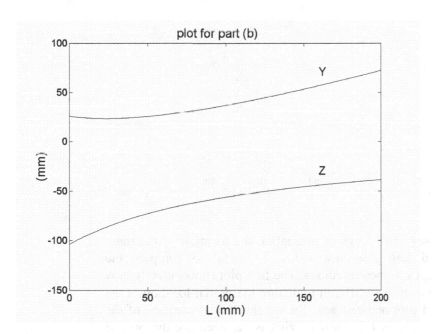

```
%%%%%%%%%%%%%% Script #2 %%%%%%%%%%%%%
% This script plots the the z coordinate
% of the center of mass versus the y
% coordinate. This is an example of a
% parametric plot.
L=0:0.5:200;
sum_mass=1.166+0.00984*L;
yb=(30+0.00492*L.^2)./sum_mass;
zb=-120.73./sum_mass;
plot(yb,zb)
xlabel('y (mm)')
ylabel('z (mm)')
title('plot for part (c)')
%%%%%%%%%%%%%% end of script %%%%%%%%%%
```

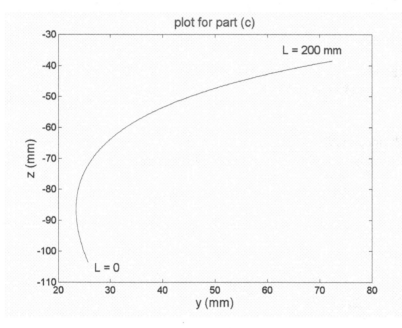

In this problem we have seen two ways of presenting the location of the mass center of the bracket-and-shaft assembly and it is useful to compare the advantages and disadvantages of these methods. The first plot shows clearly how $\overline{Y}$ and $\overline{Z}$ depend on L, however, it is difficult to picture the spatial location of the mass center. In the second plot one can actually see the spatial location of the mass center and how it moves as L is varied. What is not seen are the precise values of L at each combination of $\overline{Y}$ and $\overline{Z}$. This situation can be partially remedied by giving the values of L at the end-points of the curve, as was done in the plot above.

5.2 Problem 5/113 (Beams-External Effects)

The load per foot of beam length varies as shown. Here we would like to look at the effects of the initial load w_0 on the reactions given the constraint that the total resultant R of the distributed load w remains constant. This means that we are, in effect, studying the effects of the shape of the load. First find the required value of k in terms of w_0 and L and then (a) plot the distributed load ($w(x)$) for $w_0 = 100$, 200, and 300 lb/ft and (b) plot the reactions at the two supports as a function of w_0 for $0 \le w_0 \le 300$ lb/ft. For (a) and (b) let $R = 3500$ lb and $L = 20$ ft.

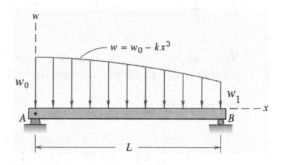

Problem Formulation

$$R = \int_0^L w\,dx = \int_0^L \left(w_0 - kx^3\right)dx = w_0 L - \frac{1}{4}kL^4$$

Solving,

$$k = \frac{4(w_0 L - R)}{L^4}$$

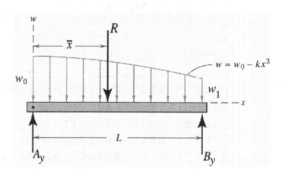

$$\bar{x} = \frac{1}{R}\int_0^L xw\,dx = \frac{1}{R}\int_0^L x\left(w_0 - kx^3\right)dx = \frac{1}{R}\left(\frac{1}{2}w_0 L^2 - \frac{1}{5}kL^5\right)$$

Substituting for k,

$$\bar{x} = \frac{L(8R - 3w_0 L)}{10R}$$

Now, from the free-body diagram,

$$\circlearrowleft \Sigma M_A = 0 = B_y L - R\bar{x}$$

$$\Sigma F_y = 0 = A_y + B_y - R$$

From which we find,

$$B_y = \frac{\bar{x}}{L}R \qquad A_y = R - B_y = \left(1 - \frac{\bar{x}}{L}\right)R$$

Needless to say, most of the substitutions indicated above are performed automatically in Maple, as we will see below.

MATLAB Scripts

```
%%%%  Script #1  %%%%%%%%%%%%%%%%%%%%%%%%
% This script plots the distributed loading
% function w(x) for w0 = 100, 200, and 300
% lb/ft
R = 3500; L = 20;
w0 = [100 200 300];
k = 4*(w0*L-R)/L^4;
x = 0:0.1:20;
w100 = w0(1)-k(1)*x.^3;
w200 = w0(2)-k(2)*x.^3;
w300 = w0(3)-k(3)*x.^3;
plot(x,w100,x,w200,x,w300)
xlabel('x (ft)')
ylabel('w(x) (lb/ft')
title('part (a), loading function w')
%%%% end of script #1 %%%%%%%%%%%%%%%%%%
```

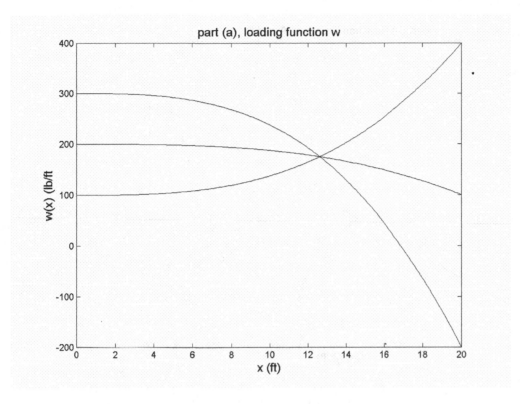

```
%%%%  Script #2  %%%%%%%%%%%%%%%%%%%%%%%%%%
% This script plots the reaction forces
% at A and B in terms of the initial loading
% function w0 given the constraint that the
% resultant of w(x) is always 3500 lb
L = 20; R = 3500;
w0 = 0:0.5:300;
xbar = L*(8*R-3*w0*L)/10/R;
B = R*xbar/L;
A = R-B;
plot(w0,A,w0,B)
xlabel('w0 (lb/ft)')
ylabel('Reaction Force (lb)')
title('part (b) Reaction Forces at A and B')
%%%% end of script #2 %%%%%%%%%%%%%%%%%%%%
```

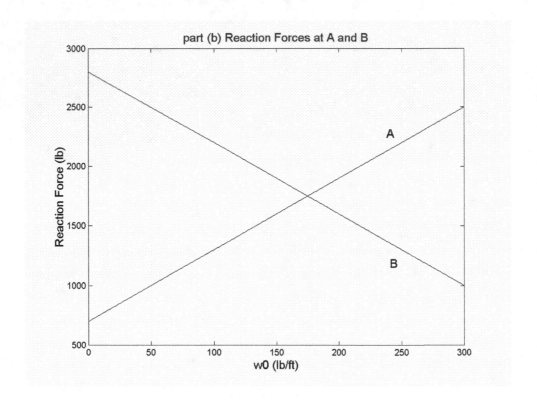

5.3 Problem 5/140 (Beams-Internal Effects)

Derive expressions for the shear force V and bending moment M as functions of x for the cantilever beam loaded as shown. For the case where $L = 10$ ft and $w_L = 400$ lb/ft, (a) plot the distributed load $w(x)$, (b) the shear force $V(x)$ and (c) the bending moment $M(x)$. In each case plot three curves for $w_0 = -400$, 0, and 400 lb/ft.

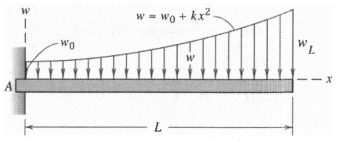

Problem Formulation

First, we need to express k in terms of w_0 and w_L. This is accomplished by substituting $x = L$ into the loading function,

$$w(x = L) = w_0 + kL^2 = w_L$$

$$k = \frac{w_L - w_0}{L^2}$$

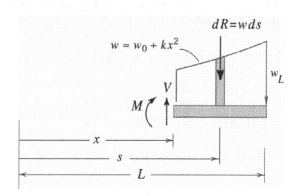

The easiest way to find $V(\mathrm{x})$ and $M(\mathrm{x})$ is from the free-body diagram shown to the right since we avoid having to find the reactions at the wall support. We do, however, have to be very careful setting up our integrals.

When we say $V(\mathrm{x})$, x is a coordinate measured positive to the right from the left end of the beam. It is essential to distinguish between that x and the dummy integration variable which will vary between x and L. Here, we denote the integration variable by s.

$$\Sigma F_y = V - \int dR \qquad V = \int_x^L w(s)\,ds = \int_x^L \left(w_0 + ks^2\right)ds$$

$$V(x) = w_0(L - x) + \frac{k}{3}\left(L^3 - x^3\right)$$

Before we sum moments first note that the moment arm of dR about a point on the cross-section at x is $s - x$,

$$\circlearrowleft \Sigma M = 0 = -M - \int (s-x) dR$$

$$M(x) = -\int_{x}^{L} (s-x) w(s) ds = -\int_{x}^{L} (s-x)\left(w_0 + ks^2\right) ds$$

$$M(x) = x w_0 (L-x) - \frac{w_0}{2}\left(L^2 - x^2\right) + \frac{xk}{3}\left(L^3 - x^3\right) - \frac{k}{4}\left(L^4 - x^4\right)$$

MATLAB Command Window and Scripts

First we will obtain the general expressions for V and M given in the problem formulation section using symbolic integration. Recall that this requires the symbolic toolbox.

EDU>> V = int('w0+k*s^2','s','x','L')

V =

w0*(L-x)+1/3*k*(L^3-x^3)

EDU>> M = -int('(s-x)*(w0+k*s^2)','s','x','L')

M =

-1/4*k*(L^4-x^4)+1/3*x*k*(L^3-x^3)-1/2*w0*(L^2-x^2)+x*w0*(L-x)

```
%%%%%%%%% script #1 %%%%%%%%%%%%%%%%%%%%%%%%%%
% This script plots the distributed loading
% function w(x) for w0 = -400, 0, and 400
% lb/ft
wL = 400; L = 10;
w0 = [-400 0 400];
k = (wL - w0)/L^2;
x = 0:0.05:10;
w1 = w0(1)+k(1)*x.^2;
w2 = w0(2)+k(2)*x.^2;
w3 = w0(3)+k(3)*x.^2;
plot(x,w1,x,w2,x,w3)
xlabel('x (ft)')
ylabel('w(x) (lb/ft')
title('part (a), loading function w')
%%%%%%%%%%%% end script %%%%%%%%%%%%%%%%%%%%%%%
```

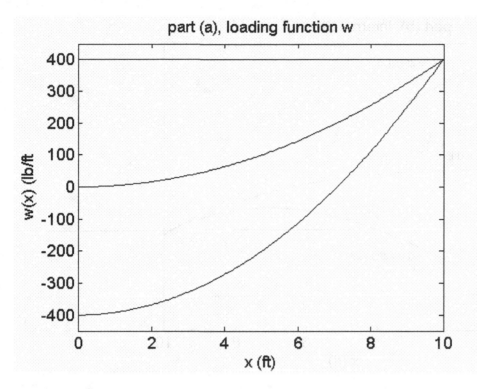

```
%%%%%%% script #2 %%%%%%%%%%%%%%%%%%%%%%%%%
% This script plots the internal shear force
%  V(x) for w0 = -400, 0, and 400 lb/ft
wL = 400; L = 10;
w0 = [-400 0 400];
k = (wL - w0)/L^2;
x = 0:0.05:10;
V1 = w0(1)*(L-x)+1/3*k(1)*(L^3-x.^3);
V2 = w0(2)*(L-x)+1/3*k(2)*(L^3-x.^3);
V3 = w0(3)*(L-x)+1/3*k(3)*(L^3-x.^3);
plot(x,V1,x,V2,x,V3)
xlabel('x (ft)')
ylabel('V(x) (lb)')
title('part (b), Internal Shear Force V')
%%%%%%%%%%% end script %%%%%%%%%%%%%%%%%%%%%
```

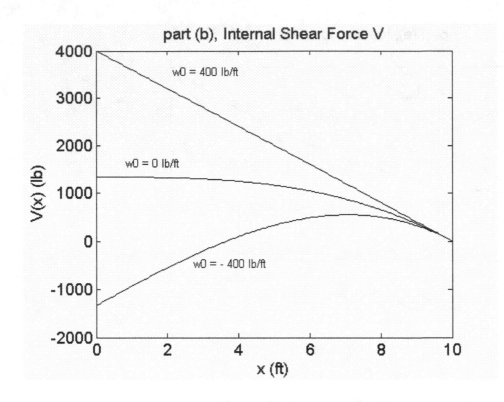

```
%%%%%%%% script #3 %%%%%%%%%%%%%%%%%%%%%%%%
% This script plots the internal moment
%  M(x) for w0 = -400, 0, and 400 lb/ft
wL = 400; L = 10;
w0 = [-400 0 400];
k = (wL - w0)/L^2;
x = 0:0.05:10;
M1 = -1/4*k(1)*(L^4-x.^4)+1/3*k(1)*x.*(L^3-x.^3)-1/2*w0(1)*(L^2-
x.^2)+w0(1)*x.*(L-x);
M2 = -1/4*k(2)*(L^4-x.^4)+1/3*k(2)*x.*(L^3-x.^3)-1/2*w0(2)*(L^2-
x.^2)+w0(2)*x.*(L-x);
M3 = -1/4*k(3)*(L^4-x.^4)+1/3*k(3)*x.*(L^3-x.^3)-1/2*w0(3)*(L^2-
x.^2)+w0(3)*x.*(L-x);
plot(x,M1,x,M2,x,M3)
xlabel('x (ft)')
ylabel('M(x) (lb-ft)')
title('part (c), Internal Moment M')
%%%%%%%%%%%% end script %%%%%%%%%%%%%%%%%%%%%%
```

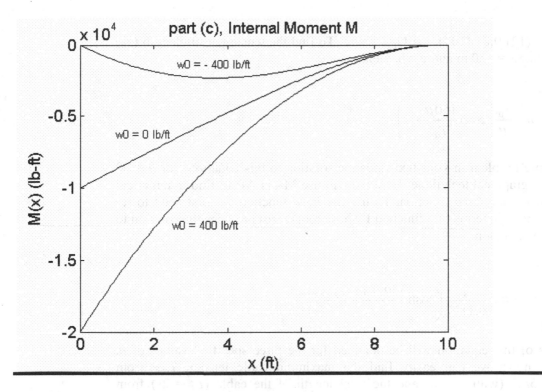

5.4 Sample Problem 5/17 (Flexible Cables)

Replace the cable of Sample Problem 5/16, which is loaded uniformly along the horizontal, by a cable which has a mass of 12 kg per meter of its own length and supports its own weight only. The cable is suspended between two points on the same level 300 m apart and has a sag of h meters (instead of the 60 m shown in the diagram. Find the tension at midlength, the maximum tension, and the total length of the cable for h = 10, 30, and 60 meters. Plot y as a function of x for these values of h.

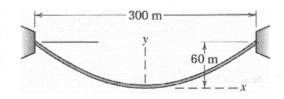

Problem Formulation

For a uniformly distributed load we have a catenary shape for the cable as described in part (c) of Article 5/8. The curve assumed by the cable ($y(x)$) is thus described by Equation 5/19

$$y = \frac{T_0}{\mu}\left(\cosh\frac{\mu x}{T_0} - 1\right)$$

where $\mu = (12)(9.81)(10^{-3}) = 0.1177$ kN/m. To find the tension at midlength (T_0) we substitute $x = 150$ m and $y = $ h, yielding

$$h = \frac{T_0}{\mu}\left(\cosh\frac{150\mu}{T_0} - 1\right)$$

The Sample Problem in your text finds the solution to this equation (for h = 60 m) using a graphical technique. Here we will use MATLAB to find a numerical solution using the *fzero* function. To use the *fzero* function we first need to re-write the above in terms of a function f whose root (zero) provides the solution to the original equation.

$$f = h - \frac{T_0}{0.1177}\left(\cosh\frac{150(0.1177)}{T_0} - 1\right)$$

The roots of this equation will be obtained for the three specified values of h. With T_0 known we can easily find the maximum cable tension T_{max} from Equation 5/22 (with y = h) and the total length of the cable ($L_c = 2s$) from Equation 5/20 (with x = 150 m).

$$T_{max} = T_0 + \mu h$$

$$L_c = 2s = \frac{2T_0}{\mu}\sinh\left(\frac{150\mu}{T_0}\right)$$

The results are:

h (m)	T_0 (kN)	T_{max} (kN)	L_c (m)
10	132.6	133.8	300.9
30	44.7	48.2	307.9
60	23.2	30.2	329.9

MATLAB Scripts and Worksheet

```
%%%%%%%%%%% Function File %%%%%%%%%%%%%%
%This function file calculates f (see problem
%formulation) which is in turn used with fzero
function f = T_0(x,h)
f = h-x./0.1177.*(cosh(0.1177*150./x)-1);
%%%%%%%%% end of function %%%%%%%%%%%%%%
```

MATLAB Command Window

```
EDU>> fzero('T_0(x,10)',100)

ans =

  132.6082

EDU>> fzero('T_0(x,30)',100)

ans =

  44.7139

EDU>> fzero('T_0(x,60)',100)

ans =

  23.1585
```
```
%%%%%%%%%%%%%% Script #1 %%%%%%%%%%%%%%%%%%%%%
% This script calculates the maximum cable
% tension Tmax and the cable length Lc given
% h and the values for T0 found numerically
% using fzero
mu = 0.1177;
h = [10 30 60];
T0 = [132.608 44.714 23.159];
Tmax = T0 + mu*h;
Lc = 2*T0/mu.*sinh(150*mu./T0);
Results(:,1)=h';
Results(:,2)=T0';
Results(:,3)=Tmax';
Results(:,4)=Lc';
Results
%%%%%%%%% end of script #1 %%%%%%%%%%%%%%%%%%%%
```

Output from script #1

Results =

```
10.0000   132.6080   133.7850   300.8871
30.0000    44.7140    48.2450   307.8560
60.0000    23.1590    30.2210   329.9142
```

```
%%%%%%%%%%%%% Script # 2 %%%%%%%%%%%%%
%This script plots y versus x using the values of T0
%found numerically using fzero.
y=inline('T0./0.1177.*(cosh(0.1177.*x./T0)-1)');
x=-150:1:150;
y10=132.61./0.1177.*(cosh(0.1177.*x./132.61)-1);
y30=44.71./0.1177.*(cosh(0.1177.*x./44.71)-1);
y60=23.16./0.1177.*(cosh(0.1177.*x./23.16)-1);
plot(x,y10,x,y30,x,y60)
xlabel('x (m)')
ylabel('y (m)')
%%%%%%%%%% end of script %%%%%%%%%%%
```

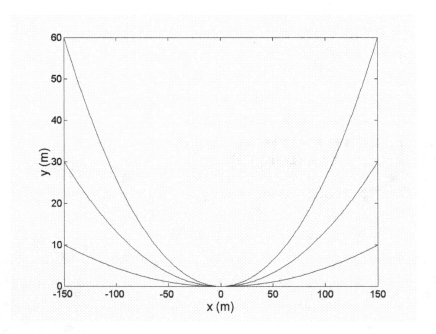

5.5 Problem 5/189 (Fluid Statics)

The rectangular gate shown in section is 10 ft long (perpendicular to the paper) and is hinged about its upper edge B. The gate divides a channel leading to a fresh-water lake on the left and a salt-water tidal basin on the right. Calculate the torque M on the shaft of the gate at B required to prevent the gate from opening in terms of h (the distance between salt water and fresh water levels). Plot M as a function of h for $0 \le h \le 6$ ft.

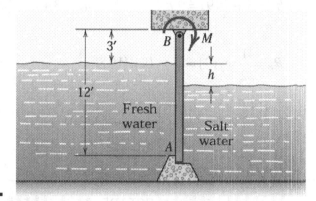

Problem Formulation

To the right is a free-body diagram for the gate. The water pressures have been replaced by linear load distributions where the maximum load intensity w is given by the general formula

$$w = \gamma dL$$

where γ is the specific weight, d is the depth of water and L is the length (into the page). Thus,

$$w_f = \gamma_f (9\,ft)(10\,ft) \quad w_s = \gamma_s (9-h)(10)$$

where $\gamma_f = 62.4 \dfrac{lb}{ft^3}$ and $\gamma_s = 64.0 \dfrac{lb}{ft^3}$.

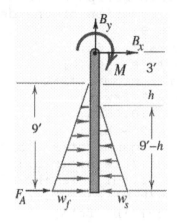

Now replace the distributed loads by their statically equivalent concentrated loads as shown on the free-body diagram to the right. Note that the contact force $F_A = 0$ since we are finding the smallest M required, i.e. the gate is on the verge of opening.

$$\circlearrowleft \Sigma M_B = 0 = P_f h_1 - P_s h_2 - M$$

$$M = P_f h_1 - P_s h_2$$

where,

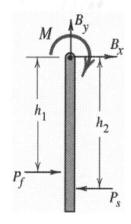

$$h_1 = 3 + \frac{2}{3}(9) = 9\,ft \qquad\qquad h_2 = 3 + h + \frac{2}{3}(9-h) = 9 + \frac{h}{3}$$

$$P_f = \frac{1}{2}w_f(9) = \frac{1}{2}\gamma_f(9)^2(10)$$

$$P_s = \frac{1}{2}w_s(9-h) = \frac{1}{2}\gamma_s(9-h)^2(10)$$

MATLAB Script

```
%%%%%%%%%%%%%% script %%%%%%%%%%%%%%%%%%
% This script plots the moment M as
% a function of the depth h
h1 = 9;
gamma_f = 62.4;
gamma_s = 64.0;
w_f = gamma_f*9*10;
Pf = 1/2*w_f*9;
h = 0:0.05:6;
h2 = 9 + h/3;
w_s = gamma_s*(9 - h)*10;
Ps = 1/2*w_s.*(9-h);
M = Pf.*h1 - Ps.*h2;
plot(h,M)
xlabel('h (ft)')
ylabel('M (lb-ft)')
title('Moment to keep the gate closed')
%%%%%%%%%%%%%%%%%%%%%%%%%%%%%%%%%%%%%%%%%%
```

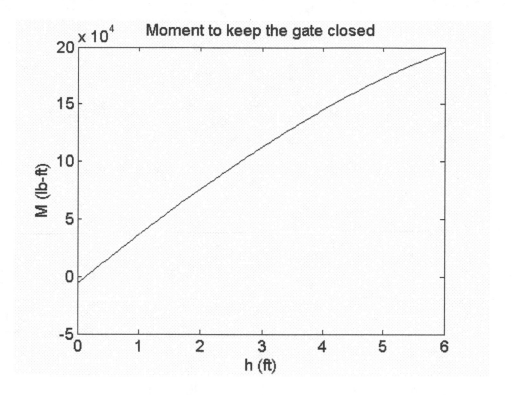

FRICTION

6

Coulomb friction (dry friction) can have a significant effect upon the analysis of engineering structures. Problem 6.1 considers three blocks stacked on an incline oriented at an arbitrary angle θ and illustrates the importance of identifying all possibilities for impending motion. In this problem there turns out to be a transition between two possibilities as the angle θ increases. Problem 6.2 considers a cylinder wedged between two rough surfaces. Three equations are solved simultaneously using the symbolic *solve* command. Problem 6.3 considers the possibility of impending slip as a person climbs to the top of a ladder. Problems 6.4, 6.5, and 6.6 are applications involving friction on wedges, screws and flexible belts.

6.1 Sample Problem 6/5 (Friction)

The three flat blocks are positioned on an incline that is oriented at an angle θ (instead of 30°) from the horizontal. Plot the maximum value of P (if no slipping occurs) versus θ. Consider only positive values of P and indicate the regions over which (1) the 50-kg block slides alone and (2) the 50-kg and 40-kg blocks slide together. The coefficients of friction μ_s are given in the figure to the right.

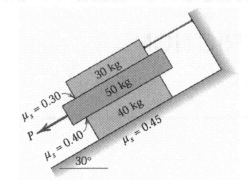

Problem Formulation

The free-body diagrams for the three blocks are shown to the right. To obtain the required plot it will be convenient to use a slightly different approach than that used in the sample problem in your text. We start by writing down the equilibrium equations without making any assumptions about where sliding occurs.

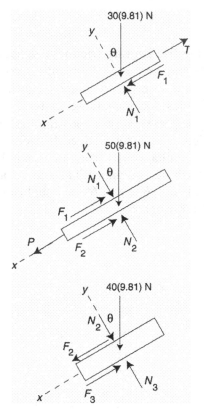

$$[\Sigma F_y = 0] \quad \text{(30-kg)} \qquad N_1 - 30(9.81)\cos\theta = 0$$
$$\text{(50-kg)} \qquad N_2 - N_1 - 50(9.81)\cos\theta = 0$$
$$\text{(40-kg)} \qquad N_3 - N_2 - 40(9.81)\cos\theta = 0$$

These equations can be readily solved for the normal forces.

$$N_1 = 30(9.81)\cos\theta \qquad N_2 = 80(9.81)\cos\theta$$
$$N_3 = 120(9.81)\cos\theta$$

$$[\Sigma F_x = 0] \quad \text{(50-kg)} \qquad P - F_1 - F_2 + 50(9.81)\sin\theta = 0$$
$$\text{(40-kg)} \qquad F_2 - F_3 + 40(9.81)\sin\theta = 0$$

Now we have two equations with four unknowns, P, F_1, F_2, and F_3. Note that we have not written the equation for the summation of forces in the x direction for the 30-kg block. The reason is that this equation introduces an additional unknown (T) that we are not interested in determining.

The next step is to make assumptions about which block(s) slide. As will be seen, either of the two possible assumptions about impending motion will reduce two of the friction forces to functions of θ only. This will result in two equations that may be solved for P and the remaining friction force. The forces calculated will be designated P_1 or P_2 to distinguish the two cases for impending slip.

Case (1): Only the 50-kg block slips.

Impending slippage at both surfaces of the 50-kg block gives $F_1 = 0.3N_1 = 88.29\cos\theta$ and $F_2 = 0.4N_1 = 313.9\cos\theta$. Substituting these results into the equilibrium equations yields

$$P_1 = 402.2\cos\theta - 490.5\sin\theta$$

Case (2): The 40 and 50-kg blocks slide together.

Impending slippage at the upper surface of the 40-kg block and lower surface of the 50-kg block gives $F_1 = 0.3N_1 = 88.29\cos\theta$ and $F_3 = 0.45N_3 = 529.7\cos\theta$. Substitution of these results into the equilibrium equations gives,

$$P_2 = 618.0\cos\theta - 882.9\sin\theta$$

Which of these two values of P represents the maximum load that can be applied without slippage on any surface is best illustrated by plotting the two expressions as a function of θ. This plot will be generated in the script below. The basic idea is that at any specified angle θ, the critical or maximum value of P will be the smaller of two values calculated.

MATLAB Script

```
theta=0:0.01:pi/4;
P1=402.2*cos(theta)-490.5*sin(theta);
P2=618*cos(theta)-882.9*sin(theta);
plot(theta*180/pi, P1, theta*180/pi, P2)
xlabel('theta (degrees)')
ylabel('P (N)')
```

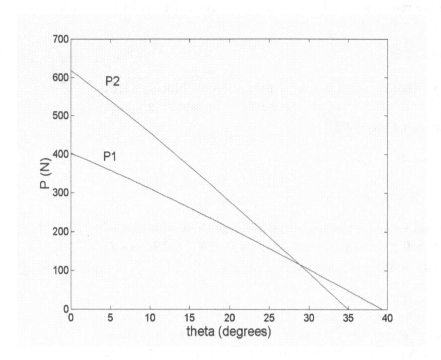

The figure above shows P_1 and P_2 plotted as a function of θ. For each θ, the critical or maximum value of P will be the smaller of two values calculated. By setting $P_1 = P_2$ we find that the two curves intersect at $\theta = 0.503$ rads (28.8°). Thus, for $\theta \leq 28.8°$ P_1 controls and the 50-kg block slides by itself while for $\theta \geq$ 28.8° P_2 controls and the 40 and 50-kg blocks slide together.

The figure below shows only the critical values for P together with an indication of the mode of slippage. This type of figure is rather tedious to produce in MATLAB so the code is not given. It is shown here for purposes of illustration.

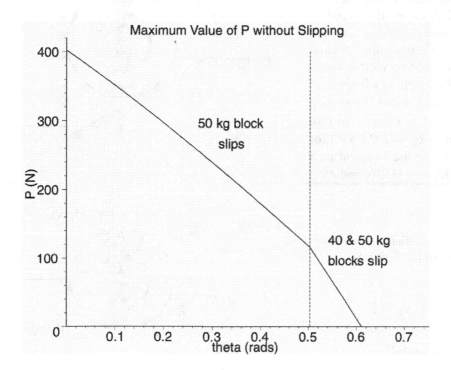

6.2 Problem 6/9 (Friction)

The 30-kg homogeneous cylinder of 400-mm diameter rests against the vertical and inclined surfaces as shown. Calculate the applied clockwise couple M which would cause the cylinder to slip. Also calculate the normal contact forces at A and B. Express your answers in terms of the angle θ and the coefficient of static friction μ_s. (a) Plot the normal forces as a function of θ ($0 \leq \theta \leq 45°$) for $\mu = 0.3$, (b) Plot the Moment as a function of θ ($0 \leq \theta \leq 45°$) for $\mu = 0.2$, 0.5, and 0.8, (c) Plot the Moment as a function of μ ($0 \leq \mu \leq 1$) for $\theta = 15$, 30, and 45°.

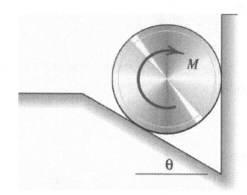

Problem Formulation

First we write the equilibrium equations from the free-body diagram for the cylinder (shown to the right).

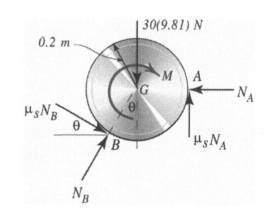

$$\circlearrowleft \Sigma M_A = 0 = M - \mu_s(N_A + N_B)0.2$$

$$\Sigma F_x = 0 = N_B \sin\theta + \mu_s N_B \cos\theta - N_A$$

$$\Sigma F_y = 0 = N_B \cos\theta - \mu_s N_B \sin\theta + \mu_s N_A - 30(9.81)$$

The last two equations can be readily solved simultaneously for N_A and N_B. Substituting these results into the first equation gives an expression that can be solved for M. The results are,

$$N_A = \frac{294.3(\sin\theta + \mu\cos\theta)}{\cos\theta(1 + \mu^2)} \qquad N_B = \frac{294.3}{\cos\theta(1 + \mu^2)}$$

$$M = \frac{58.86\mu(1 + \sin\theta + \mu\cos\theta)}{\cos\theta(1 + \mu^2)}$$

MATLAB Scripts

```
%%%%%%%%%%%%%%% Script #1 %%%%%%%%%%%%%%%%%%%%
% This script solves the three equilibrium equations
% simultaneously for NA (x), NB (y) and M (z)
eqn1='z-mu*(x+y)*.2=0'
eqn2='y*sin(theta)+mu*y*cos(theta)-x=0'
eqn3='y*cos(theta)-mu*y*sin(theta)+mu*x-30*9.81'
[x,y,z] = solve(eqn1,eqn2,eqn3)
%%%%%%%%% end of script %%%%%%%%%%%%%%%%%%%%%%%
```

Executing this script gives:

eqn1 =
z-mu*(x+y)*.2=0

eqn2 =
y*sin(theta)+mu*y*cos(theta)-x=0

eqn3 =
y*cos(theta)-mu*y*sin(theta)+mu*x-30*9.81

x =
294.3*(sin(theta)+mu*cos(theta))/cos(theta)/(1.+mu^2)

y =
294.3/cos(theta)/(1.+mu^2)

z =
58.86*mu*(sin(theta)+mu*cos(theta)+1.)/cos(theta)/(1.+mu^2)

```
%%%%%%% script #2 (part a) %%%%%%%%%%%%%%%%%%%%%%%%%%
% this script plots the normal forces as functions
% of theta
mu = 0.3;
th = 0:0.1:45;
theta = th*pi/180;
NA = 294.3*(sin(theta)+mu*cos(theta))./cos(theta)/(1.+mu^2);
NB = 294.3./cos(theta)/(1.+mu^2);
plot(th,NA,th,NB)
xlabel('theta (degrees)')
ylabel('Force (N)')
title('Part (a) Normal Forces')
%%%%%%%%%%%%% end of script %%%%%%%%%%%%%%%%%%%%%%%%%
```

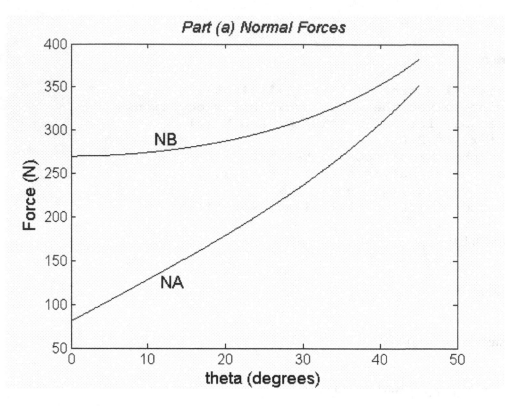

```
%%%%%%% script #3 (part b) %%%%%%%%%%%%%%%%%%%%%%%%%%%
% this script plots the Moment as a function
% of theta for mu = .2, .5, and .8
M = inline('58.86*mu*(sin(x)+mu*cos(x)+1)./cos(x)/(1+mu^2)')
th = 0:0.1:45;
theta = th*pi/180;
M1 = M(.2,theta);
M2 = M(.5,theta);
M3 = M(.8,theta);
plot(th,M1,th,M2,th,M3)
xlabel('theta (degrees)')
ylabel('Moment (N-m)')
title('Part (b) Moment')
%%%%%%%%%%%%%% end of script %%%%%%%%%%%%%%%%%%%%%%%%%%
```

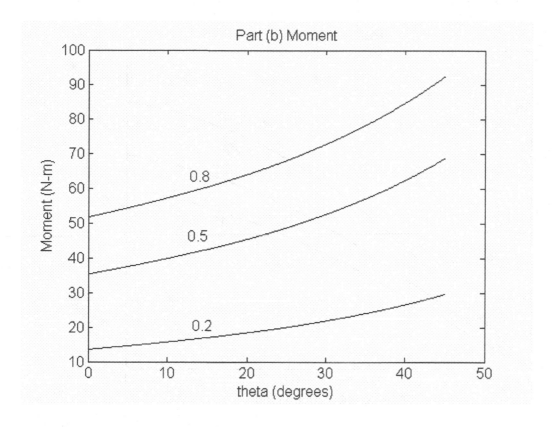

```
%%%%%%% script #3 (part c) %%%%%%%%%%%%%%%%%%%%%%%%%%%%
% this script plots the Moment as a function
% of mu for theta = 15, 30, and 45 degrees
M = inline('58.86*mu.*(sin(x)+mu*cos(x)+1)./cos(x)./(1+mu.^2)')
mu = 0:0.01:1;
M1 = M(mu,15*pi/180);
M2 = M(mu,30*pi/180);
M3 = M(mu,45*pi/180);
plot(mu,M1,mu,M2,mu,M3)
xlabel('coefficient of friction mu')
ylabel('Moment (N-m)')
title('Part (c) Moment')
%%%%%%%%%%%%% end of script %%%%%%%%%%%%%%%%%%%%%%%%%%
```

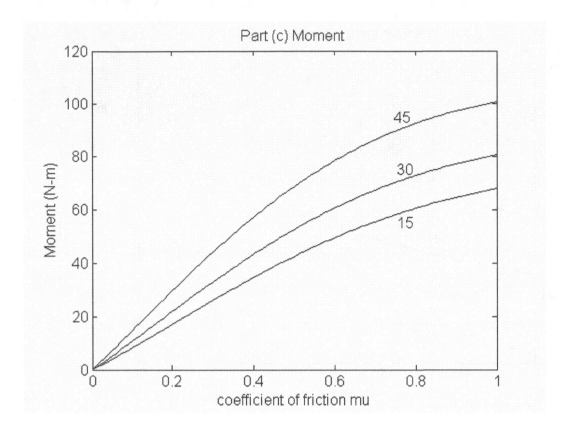

6.3 Problem 6/44 (Friction)

The 15-kg ladder is 4-m long and is inclined at an arbitrary angle θ. The top of the ladder has a small roller, and at the ground the coefficient of friction is μ. Determine the distance s (in terms of θ and μ) to which the 90-kg painter can climb without causing the ladder to slip at its lower end A. Plot s as a function of θ ($0 \le \theta \le 90°$) for $\mu = 0.2$, 0.4, and 0.6. Limit the vertical (s) axis to be from 0 to 5-m. Provide a physical explanation for the angles θ where (a) $s = 0$ and (b) $s \ge 4$-m. The mass center of the painter is directly above his feet.

Problem Formulation

First we write the equilibrium equations from the free-body diagram for the ladder (shown to the right).

$$\Sigma F_x = 0 = \mu N_A - N_B$$
$$\Sigma F_y = 0 = N_A - 90g - 15g$$
$$\circlearrowleft \Sigma M_A = 0 = 90g(s\cos\theta) + 15g(2\cos\theta) - N_B(4\sin\theta)$$

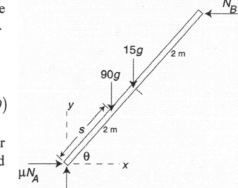

The first two equations can be readily solved for $N_B = 105\mu g$. Substituting this result into the third equation gives an expression that can be solved for s.

$$s = \frac{14\mu\sin\theta - \cos\theta}{3\cos\theta}$$

MATLAB Scripts

```
%%%%%%%%%% Script #1 %%%%%%%%%%%%%%
%This program solves for s symbolically from the moment
%equilibrium equation after substituting for NB.
f='90*g*s*cos(theta)+30*g*cos(theta)-105*mu*g*4*sin(theta)=0'
s=solve(f,'s')
%%%%%%%%%% end of script %%%%%%%%%%
```

Output of script # 1

f = 90*g*s*cos(theta)+30*g*cos(theta)-105*mu*g*4*sin(theta)=0

s = -1/3*(cos(theta)-14*mu*sin(theta))/cos(theta)

```
%%%%%%%%%% Script #2 %%%%%%%%%%%%%%
%This script plots x as a function of theta for
%the specified values of mu.
s=inline('-1/3*(cos(th)-14*mu*sin(th))./cos(th)')
th=0:0.01:pi/2;
plot(th*180/pi,s(.2,th),th*180/pi,s(.4,th),th*180/pi,s(.6,th))
axis([0 90 0 5])
xlabel('theta (degs)')
ylabel('s (m)')
%%%%%%%%%% end of script %%%%%%%%%%
```

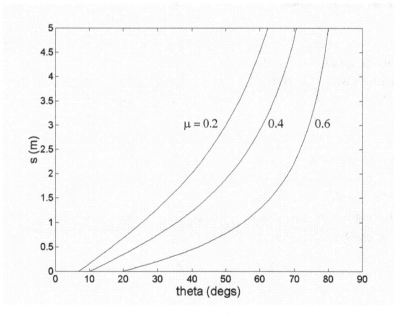

The physical interpretation is as follows. The value of θ for which $s = 0$ corresponds to the orientation where the ladder is on the verge of slipping before the painter steps on it. Obviously, the ladder will slip immediately if she does try to step on it. The values of θ for which $s \geq 4$ correspond to orientations where the painter can climb to the top of the ladder without having it slip.

6.4 Sample Problem 6/6 (Friction on Wedges)

The horizontal position of the 500-kg rectangular block of concrete is adjusted by the wedge under the action of the force **P**. Let the wedge angle be θ (instead of 5°). Also let μ_1 and μ_2 be the coefficient of static friction at the two wedge surfaces and between the block and the horizontal surface respectively. (a) Derive a general expression for P (the least force required to move the block) in terms of θ, μ_1 and μ_2. (b) For $\mu_2 = 0.6$, plot P as a function of μ_1 for $\theta = 5$, 15, and 25°. (c) For $\theta = 5°$, plot P as a function of μ_1 for $\mu_2 = 0.2$, 0.4, 0.6, and 0.8.

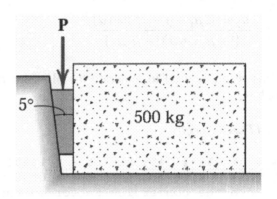

Problem Formulation

(a) First we write the equilibrium equations from the free-body diagrams for the wedge and for the block.

For the Block

$$\Sigma F_x = 0 = N_2 - \mu_2 N_3 \qquad \Sigma F_y = 0 = N_3 - mg - \mu_1 N_2$$

These two equations are readily solved for N_2 and N_3.

$$N_2 = \frac{\mu_2 mg}{1 - \mu_1 \mu_2} \qquad\qquad N_3 = \frac{mg}{1 - \mu_1 \mu_2}$$

For the Wedge

$$\Sigma F_x = 0 = N_1 \cos\theta - \mu_1 N_1 \sin\theta - N_2$$

$$\Sigma F_y = 0 = N_1 \sin\theta + \mu_1 N_1 \cos\theta + \mu_1 N_2 - P$$

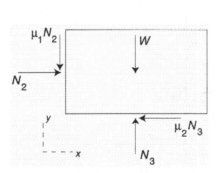

After substituting for N_2 we solve the first equation for N_1. Substituting this result into the second yields an expression for P.

$$P = \frac{\mu_2 mg\left(2\mu_1 + \left(1 - \mu_1^2\right)\tan\theta\right)}{\left(1 - \mu_1 \tan\theta\right)\left(1 - \mu_1\mu_2\right)}$$

This result is used to generate the plots for parts **(b)** and **(c)** in the worksheet below.

MATLAB Scripts

```
%%%%%%%%%%%%%%%%%% Script # 1 %%%%%%%%%%%%%%%%%
% This script produces the plot for part (b)
% Note the conversion factor on theta in the following
% function. This allows us to use degrees rsther than
% radians when we use P. Also note the division by 1000,
% converting to kN.
P=inline('.6*500*9.81*(tan(th*pi/180)*(1-mul^2)+2*mul)/
(1-mul*tan(th*pi/180))/(1-mul*.6)/1000');
P=vectorize(P);
mul=0:0.01:0.8;
plot(mul,P(mul,5),mul,P(mul,15),mul,P(mul,25))
xlabel('mu_1')
ylabel('P (kN)')
```

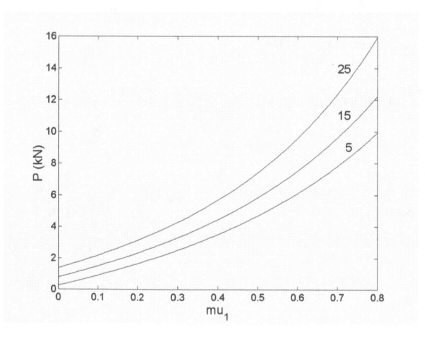

```
%%%%%%%%%%%%%%%%% Script # 2 %%%%%%%%%%%%%%%
% This script produces the plot for part (c)
% Note that P is a function of mu1 and mu2 in this case.
%Also note the division by 1000, converting to kN.
P=inline('mu2*500*9.81*(tan(5*pi/180)*(1-mu1^2)+2*mu1)/(1-
mu1*tan(5*pi/180))/(1-mu1*mu2)/1000');
P=vectorize(P);
mu1=0:0.01:0.8;
plot(mu1,P(mu1,.2),mu1,P(mu1,.4),mu1,P(mu1,.6),mu1,P(mu1,.8))
xlabel('mu_1')
ylabel('P (kN)')
```

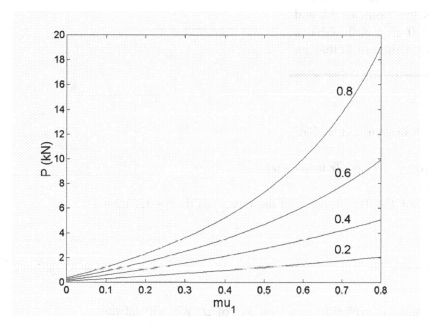

6.5 Problem 6/60 (Friction on Screws)

The large turnbuckle supports a cable tension of 10,000 lb. The 1.25 in. screws have a mean diameter of 1.150 in. and have five square threads per inch. The coefficient of friction for the threads is μ. Both screws have single threads and are prevented from turning. Determine the moments M_T and M_L that must be applied to the body of the turnbuckle in order to tighten and loosen it respectively. Plot the moments M_T and M_L as functions of μ for $0 \le \mu \le 1$. Give a physical explanation for any values of M that are less than zero.

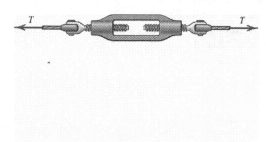

Problem Formulation

From equations 6/3 and 6/3b in your text we have,

$$M_T = 2Tr\tan(\phi + \alpha) \qquad M_L = 2Tr\tan(\phi - \alpha)$$

where $T = 10,000$ lb, the lead $L = 1/5$ in./rev. and the mean radius $r = 1.15/2 = 0.575$ in. Also,

$$\alpha = \tan^{-1}\frac{L}{2\pi r} \qquad \phi = \tan^{-1}\mu$$

Substitution will give M_T and M_L explicitly as a function of μ. We will let the computer substitute for us.

MATLAB script

```
T=10000; L=1/5; r=1.15/2;
alpha=atan(L/2/pi/r);
mu=0:0.01:1;
phi=atan(mu);
M_T=2*T*r*tan(alpha+phi);
M_L=2*T*r*tan(phi-alpha);
M=0*mu; % used to place a horizontal line at x=0
plot(mu,M_T,mu,M_L,mu,M)
xlabel('coefficient of friction')
ylabel('Moment (lb-in)')
```

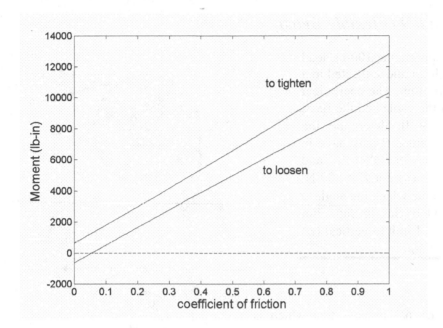

Note that the predicted values for M_L are negative for small μ. What is the physical explanation for this? You may recall that the equation for M_L derived in your text assumes that $\phi > \alpha$. When $\phi = \alpha$, the turnbuckle will be on the verge of loosening without an external moment applied to its body. The condition $\phi = \alpha$ gives, from the equations above,

$$\mu = \frac{L}{2\pi r} = \frac{1/5}{2\pi(0.575)} = 0.00553$$

This is the value of μ for which $M_L = 0$ in the plot above. We conclude then that the turnbuckle will loosen without any external moment being applied whenever $\mu < 0.00553$.

6.6 Sample Problem 6/9 (Flexible Belts)

A flexible cable which supports the 100-kg load is passed over a circular drum and subjected to a force P to maintain equilibrium. The coefficient of static friction between the cable and the fixed drum is μ. (a) For $\alpha = 0$, determine the maximum and minimum values P may have in order not to raise or lower the load. Plot $P_{\max}$ and $P_{\min}$ versus μ for $0 \le \mu \le 1$. (b) For $P = 500$ N, determine the minimum value which the angle α may have before the load begins to slip. Plot $\alpha_{\min}$ versus μ for $0 \le \mu \le 1$. Limit to vertical (α) axis to be between −60 and 360°.

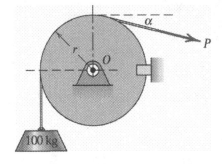

Problem Formulation

Here we will use Equation 6/7 in your text which is repeated below for convenience.

$$T_2 = T_1 e^{\mu\beta}$$

Recall that in deriving this formula it was assumed that $T_2 > T_1$.

981 N
(a) $\alpha = 0$, $\beta = \pi/2$

(a) With $\alpha = 0$ the contact angle is $\beta = \pi/2$ rad. For impending upward motion of the load we have $T_2 = P_{\max}$ and $T_1 = 981$ N. Thus

$$P_{\max} = 981 e^{\mu\pi/2}$$

For impending downward motion of the load we have $T_2 = 981$ N and $T_1 = P_{\min}$.

981 N
(b) $P = 500$ N, $\beta = \alpha + \pi/2$

$$981 = P_{\min} e^{\mu\pi/2} \quad \text{or} \quad P_{\min} = 981 e^{-\mu\pi/2}$$

(b) With $P = 500$ N we have $\beta = \pi/2 + \alpha$, $T_2 = 981$ N and $T_1 = P = 500$ N. From Equation 6/7 we have,

$$981/500 = e^{\mu(\pi/2+\alpha)}$$

Taking the natural log of both sides of the above equation and solving for α gives,

$$\alpha = \frac{\ln(981/500)}{\mu} - \frac{\pi}{2}$$

MATLAB Scripts

```
%%%%%%%%%%%%% Script #1 %%%%%%%%%%%%%%%%%
% This script plots Pmax and Pmin versus
% the friction coefficient mu
mu=0:0.005:1;
Pmax=981*exp(mu*pi/2)/1000; % note conversion to kN
Pmin=981*exp(-mu*pi/2)/1000;
plot(mu,Pmax,mu,Pmin)
xlabel('coefficient of friction')
ylabel('P (kN)')
```

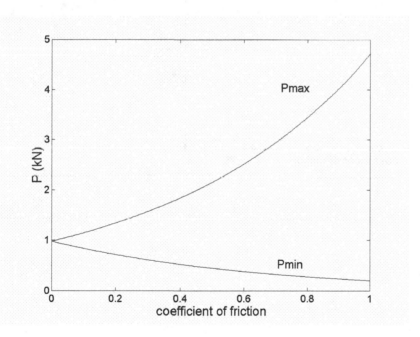

```
%%%%%%%%%%%%% Script #2 %%%%%%%%%%%%%%%%%
% This script plots the minimum angle alpha
% versus the friction coefficient mu
mu=0:0.005:1;
alpha=log(981/500)./mu-pi/2
plot(mu,alpha*180/pi)
xlabel('coefficient of friction')
ylabel('alpha (degs)')
axis([0 1 -60 360])
```

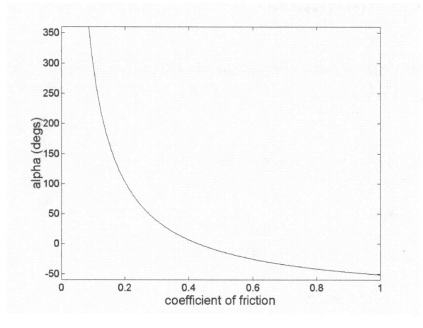

VIRTUAL WORK

7

This chapter considers the application of the principle of virtual work to the equilibrium and stability analysis of engineering structures. Problem 6.1 looks at the compressive force developed in a hydraulic cylinder that is part of a mechanism for elevating automobiles. The results of a parametric study suggest an obvious improvement in the design of the system. The formulation of problem 7.2 results in an equation that cannot be solved exactly. When this occurs one generally has to make a choice between a graphical or numerical solution. This problem illustrates in some detail a general graphical approach that is very useful in some situations. For purposes of comparison, a numerical solution is also obtained with Maple's *fsolve* function. Problem 7.3 takes a look at a problem with multiple equilibrium states and evaluates these in terms of their stability.

7.1 Problem 7/30 (Virtual Work, Equilibrium)

Express the compression C in the hydraulic cylinder of the car hoist in terms of the angle θ. The mass of the hoist is negligible compared to the mass m of the vehicle. (a) Plot the non-dimensional compression C/mg as a function of θ for three non-dimensional length ratios; $b/L =$ 0.1, 0.5, and 0.9. Let θ range between 0 and 90° and limit the vertical scale to between 0 and 3. (b) Plot C/mg as a function of b/L for several values of θ letting b/L vary between 0 and 2. From your results, see if you can improve the design of the hoist system by re-positioning the hydraulic cylinder.

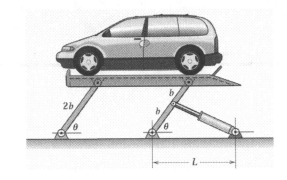

Problem Formulation

The active force diagram for the system is shown to the right. Let the length of the cylinder (AB in the diagram) be a so that the virtual work done by C is $C\delta a$. The principle of virtual work applied to this system gives,

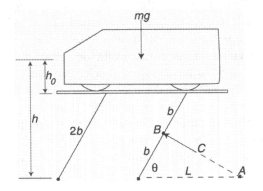

$$\delta U = 0 = C\delta a - mg\delta h$$

Now we have to do some geometry to relate the virtual displacements δa and δh to the angle θ. From the diagram we can write a^2 as follows.

$$a^2 = (b\sin\theta)^2 + (L - b\cos\theta)^2$$

Note that it is not actually necessary to solve for a in order to determine the virtual displacement δa. All you need is an expression relating a and θ. In the present case it is somewhat easier to find the variation of the expression on the right hand side of the equation above than it would be to find the variation of the square root of that expression. Taking the variation of the above equation yields,

$$2a\delta a = 2b\sin\theta(b\cos\theta\delta\theta) + 2(L - b\cos\theta)(b\sin\theta\delta\theta)$$

Solving for δa and simplifying yields,

$$\delta a = \frac{Lb}{a}\sin\theta\delta\theta$$

where $\quad a = \sqrt{(b\sin\theta)^2 + (L - b\cos\theta)^2} = L\sqrt{1 + \left(\frac{b}{L}\right)^2 - 2\frac{b}{L}\cos\theta}$

Now we need to find a relationship between δh and θ.

$$h = 2b\sin\theta + h_0 \qquad \delta h = 2b\cos\theta\delta\theta$$

Now we substitute the virtual displacements δa and δh into the virtual work equation above.

$$C\left(\frac{Lb}{a}\sin\theta\right)\delta\theta - mg(2b\cos\theta)\delta\theta = 0$$

$$C = \frac{2mga}{L}\cot\theta$$

Solving for the non-dimensional compression C/mg and substituting for a yields,

$$\frac{C}{mg} = 2\cot\theta\sqrt{1 + \left(\frac{b}{L}\right)^2 - 2\frac{b}{L}\cos\theta}$$

Before generating the required plots it might be useful to introduce some non-dimensional parameters. Letting $C' = C/mg$ and $\eta = b/L$ we can re-write the above equation as

$$C' = 2\cot\theta\sqrt{1 + \eta^2 - 2\eta\cos\theta}$$

Now we need to plot C' as a function of θ for $\eta = 0.1$, 0.5, and 0.9.

MATLAB Scripts

```
%%%%%%%%%%% Script # 1 %%%%%%%%%%
% This script generates the plot for part (a)
C=inline('2*cot(theta).*sqrt(1+eta.^2-2*eta.*cos(theta))')
theta=0:0.01:pi/2;
plot(theta,C(0.1,theta),theta,C(0.5,theta),theta,C(0.9,theta))
axis([0,pi/2,0,3])
xlabel('theta (rads)')
ylabel('C/mg')
```

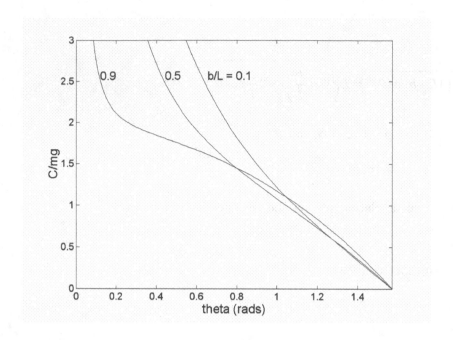

```
%%%%%%%%%%%% Script # 2 %%%%%%%%%%
% This script generates the plot for part (b)
C=inline('2*cot(theta).*sqrt(1+eta.^2-2*eta.*cos(theta))')
eta=0:0.01:2;
plot(eta,C(eta,10*pi/180),eta,C(eta,30*pi/180))
hold on
plot(eta,C(eta,50*pi/180),eta,C(eta,70*pi/180))
axis([0,2,0,3])
xlabel('b/L')
ylabel('C/mg')
hold off
```

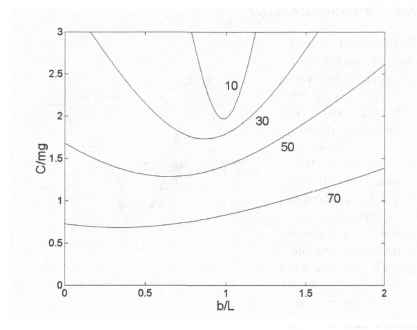

The first diagram (part (a)) shows that the present design has some real problems for small angles θ. To appreciate this, remember that we are plotting the non-dimensional force. If the non-dimensional force equals 3, for example, the compressive force in the cylinder will be three times the weight of the vehicle. The main intent of the diagram for part (b) is to show that changing the design by fine-tuning b/L will not really remove the real problem.

So, what is the physical reason for the compressive force approaching infinity as θ approaches zero? Look again at the active force diagram and try to visualize how the orientation of C changes as θ gets smaller. As θ gets smaller, the vertical component of C gets smaller, making it harder and harder for the cylinder to support the weight of the vehicle. Without changing the overall design too much, the most obvious thing to do is to recess the cylinder somewhat so that it never approaches a horizontal orientation. In other words, move point A (the base of the cylinder) vertically downward. This is illustrated by the modified drawing to the right. Note that, with this arrangement, the vertical component of C will not approach zero at small θ.

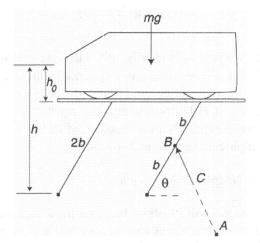

7.2 Sample Problem 7/5 (Potential Energy)

The two uniform links, each of mass m, are in the vertical plane and are connected and constrained as shown. As the angle θ between the links increases with the application of the horizontal force P, the light rod, which is connected at A and passes through a pivoted collar at B, compresses the spring of stiffness k. If the spring is uncompressed in the position where $\theta = 0$, determine the force P which will produce equilibrium at the angle θ. (a) Develop a graphical approach for determining the equilibrium value of θ corresponding to a given force P. Use a computer to generate one or more plots that illustrate your approach. (b) Obtain a numerical solution for the equilibrium angle θ for the case $P = 100$ lb, $mg = 50$ lb, $k = 20$ lb/in, and $b = 5$ in.

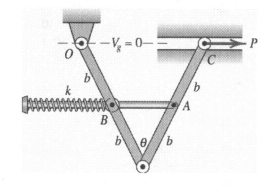

Problem Formulation

The first part of this problem (determining the force P which will produce equilibrium at the angle θ) is identical to the sample problem in your text. Therefore, we start with the equation obtained in the sample problem. Be sure that you understand how this equation was derived.

$$P = kb \sin \frac{\theta}{2} + \frac{1}{2} mg \tan \frac{\theta}{2}$$

This equation is ideally suited to situations where you would like to find the force P that will result in some specified equilibrium angle θ. Here we are interested in the opposite situation. Given a force P we would like to calculate the equilibrium value of θ. Unfortunately, the equation above cannot be inverted analytically to give θ explicitly as a function of P. In situations like this you should try either a graphical or numerical solution.

(a) Graphical Approach

Here we will illustrate two graphical approaches. The first is suggested at the end of the sample problem in your text and is suitable for solving specific problems. The second approach is more general and is suitable for certain types of design situations.

(a1) *Specific Graphical Approach*

The specific approach is suitable for situations where P, b, m and k are all specified. For example, consider the case where you want the equilibrium angle θ for only one situation. For convenience, we'll take the specific case defined in part (b).

$$P = 100 \text{ lb}, mg = 50 \text{ lb}, k = 20 \text{ lb/in, and } b = 5 \text{ in.}$$

Here we forget for the moment that P is given and plot, from the expression above, P as a function of θ. Superimposed on this plot would be a horizontal line at $P = 100$. This horizontal line can either be drawn at the same time the plot is generated or by hand after the plot has been printed. After the plot has been printed you would next find the intersection between the horizontal line and the curve. Since the curve represents the solution for all θ, the intersection will be the solution for the specific case where $P = 100$ lb. The equilibrium value of θ can thus be found by dropping a vertical line from the point of intersection to the θ axis. The procedure is illustrated in the figure to the right. From this plot we can estimate the equilibrium angle to be about $\theta = 95°$. For purposes of comparison, the numerical approach in part (b) gives $\theta = 94.05°$.

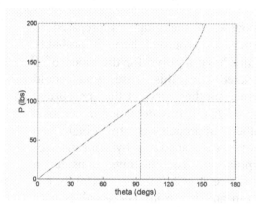

(a2) *General Graphical Approach*

What we have in mind here are certain types of design situations where specific values of parameters have not yet been decided. Here we want to develop a graphical approach that would not require us to run to a computer each time a specific case is considered. This is what is meant by a general graphical approach. The effectiveness and generality of graphical solutions can be significantly increased by first reducing, as much as possible, the number of parameters that appear in the equation. In the present case we can reduce the number of parameters from three (k, b, and m) to one by dividing both sides of the equation by the weight mg. Since P, kb and mg all have units of force, this operation results in a non-dimensional equation relating a non-dimensional force $P' = P/mg$ to the equilibrium angle θ. The only parameter remaining is the non-dimensional quantity $kb/mg = \beta$.

$$P' = \frac{P}{mg} = \beta \sin\frac{\theta}{2} + \frac{1}{2}\tan\frac{\theta}{2} \quad \text{where } \beta = \frac{kb}{mg}.$$

The plot to the right shows P' versus θ for several, evenly spaced, values of β. For a specific problem you would be asked to find θ given P, k, b, and m. One would first determine P' and β and then construct a horizontal line at the calculated value of P' and find where that line intersects a curve at the appropriate value of β. Dropping a vertical line from this intersection gives the angle θ.

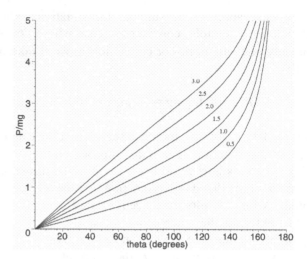

Of course, if the graph is prepared in advance, as proposed here, it probably will not have a curve for the value of β calculated. In this case you could generate another curve at the appropriate value of β, but this defeats our purpose. Another approach is to obtain an approximate answer by means of interpolation. To illustrate, consider a case where $P = 220$ lb, $mg = 100$ lb, $k = 29$ lb/in and $b = 6$ in. These values give $P' = 2.2$ and $\beta = 1.74$. The second plot to the right illustrates the interpolation procedure. Start with the horizontal line at $P' = 2.2$ and then follow this line to where it reaches a point about halfway between the curves for $\beta = 1.5$ and 2. Dropping a vertical line from this point gives a value for the equilibrium angle θ of about 112°.

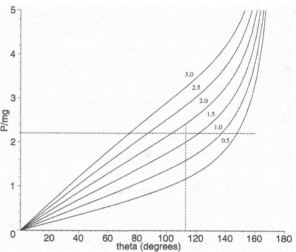

The MATLAB scripts below will show how to generate plots such as those shown above. Of course, the construction of vertical lines and interpolation are operations that are performed after the curves are generated and printed.

(b) The numerical solution is obtained with the fzero function in a worksheet below.

MATLAB Scripts

```
%%%%%%%% Script #1 (for part (a1) above %%%%%%%%%
% This script plots P as a function of theta for the
% special case mg =50 lb, k = 20 lb/in and b = 5 inches
% A horizontal line is drawn at P = 100 lb in order to
% determine the equilibrium angle theta for that load.
k=20; mg=50; b=5;
theta=0:0.01:pi;
P=k*b*sin(theta/2)+1/2*mg*tan(theta/2);
Ph=100; % for plotting a horizontal line
plot(theta*180/pi,P,theta*180/pi,Ph)
xlabel('theta (degs)')
ylabel('P (lbs)')
axis([0,180,0,200])
```

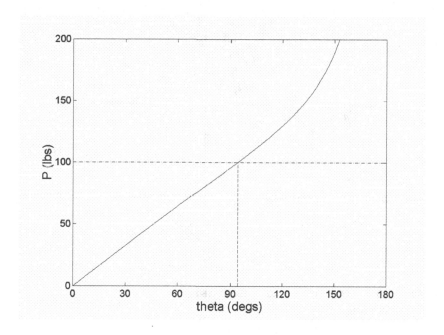

The vertical line was added with the plot editor after running the script above.

```
%%%%%%%%% Script #2 (for part (a2) above %%%%%%%%%%%%%
% This script plots the non-dimensional force P' as a function
% of theta for several values of the non-dimensional parameter
% beta. This plot is used for the general graphical approach in
% part (a2) above
P=inline('beta*sin(theta/2)+1/2*tan(theta/2)')
theta=0:0.01:pi;
plot(theta*180/pi,P(3,theta),theta*180/pi,P(2.5,theta))
hold on
plot(theta*180/pi,P(2,theta),theta*180/pi,P(1.5,theta))
plot(theta*180/pi,P(1,theta),theta*180/pi,P(0.5,theta))
xlabel('theta (degs)')
ylabel('P/mg')
axis([0,180,0,5])
hold off
```

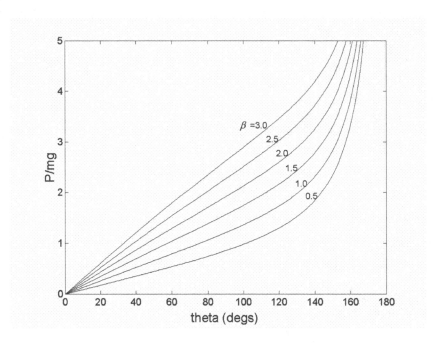

Part (b)

Here we will use fzero to find a numerical solution for θ given P = 100 lb, mg = 50 lb, k = 20 lb/in and b = 5 inches. Remember that fzero finds the zeros of an equation. For this reason we re-write the first equation in the formulation section above so that its root will be the solution to our problem. The equation is,

$$(20)(5)\sin\frac{\theta}{2} + \frac{1}{2}(50)\tan\frac{\theta}{2} - 100$$

This equation is placed in a function m-file so that it can be called with the fzero function. The name of the function file is P100.m.

```
%%%%%%%% Script #3 (a function m-file for part (b)%%%%%%%%
function y = P100(x)
y=100*sin(x/2)+25*tan(x/2)-100;
%%%%%%%%% end of m-file %%%%%%%%%%%%%%%%%%%%%%%%%%%%%%%%%%%%%%
```

Now we can call the function from a MATLAB worksheet.

EDU» fzero('P100',1)
Zero found in the interval: [0.094903, 1.9051].
ans = 1.6415

Thus, θ = 1.6415 rads (94.05°). One of the points here is, of course, that it is very difficult to beat the numerical approach for accuracy and ease of application.

7.3 Problem 7/79 (Potential Energy and Stability)

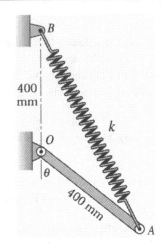

The uniform link OA has a mass of 20 kg and is supported in the vertical plane by the spring AB whose unstretched length is 400 mm. Plot the total potential energy V as a function of θ from $\theta = -15°$ to $\theta = 120°$. Consider two cases, $k = 100$ and 900 N/m. Determine all equilibrium positions (angles θ) and their stability for each case. Take $Vg = 0$ on a level through O.

Problem Formulation

Imagine dividing triangle OAB into two identical right triangles to note that half of the length AB is $0.4\sin\big((180-\theta)/2\big)$. Also note the trig identity $\sin\big((180-\theta)/2\big) = \cos(\theta/2)$. Thus,

$$AB = 2(0.4)\sin\left(\frac{180-\theta}{2}\right) = 0.8\cos\frac{\theta}{2}$$

Since the unstretched length of the spring is 0.4 m, the elastic potential energy for the springs is,

$$V_e = \frac{1}{2}k(AB-0.4)^2 = \frac{1}{2}k\left(0.8\cos\frac{\theta}{2} - .4\right)^2$$

Taking $V_g = 0$ on the horizontal plane through O we obtain the gravitational potential energy of the system as,

$$V_g = 20(9.81)(-0.2\cos\theta) = -39.24\cos\theta$$

Thus, the total potential energy and its derivative are,

$$V = V_g + V_e = \frac{1}{2}k\left(0.8\cos\frac{\theta}{2} - 0.4\right)^2 - 39.24\cos\theta$$

$$\frac{dV}{d\theta} = 39.24\sin\theta - 0.4k\left(0.8\cos\frac{\theta}{2} - 0.4\right)\sin\frac{\theta}{2}$$

The equilibrium positions and their stability will be determined below.

MATLAB Script

```
%%%%%%%%% Script #1 %%%%%%%%%%%%%
% This script plots the potential energy V
% for k = 100 and 900 N/m
V=inline('1/2*k*(0.8*cos(x/2)-.4).^2-39.24*cos(x)')
theta=-15*pi/180:0.01:120*pi/180;
plot(theta,V(100,theta),theta,V(900,theta))
xlabel('theta (rads)')
ylabel('V (J)')
%%%%%%%%%%%%%%%%%%%%%%%%%%%%%%%%%%%%%%
```

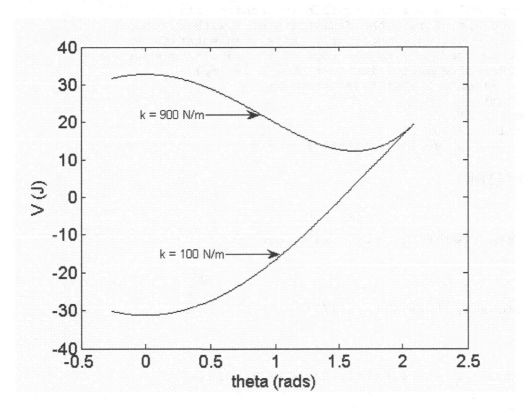

Equilibrium positions and their stability.

Recall from your text that stable equilibrium occurs at positions where the total potential energy of the system is a minimum. Unstable equilibrium, on the other hand, occurs at positions where the total potential energy of the system is a maximum. From this general principle we see that the spring with stiffness 100 N/m has only one equilibrium position at $\theta = 0$ and it is stable. The spring with stiffness 900 N/m has two equilibrium positions, an unstable one at $\theta = 0$ and a stable one just beyond 1.5 radians. This second position can be determined more precisely by finding the angles for which $dV/d\theta = 0$. This is carried out in the following.

```
%%%%%%%%% Script #2 %%%%%%%%%%%%%
% This script solves for the equilibrium configurations
% for k = 900 N/m. First, the derivative of V with respect
% to theta is evaluted symbolically. k is then substituted
% and solve is used to find the root of the resulting equation.
% Note in the solve expression that dV = 0 is implied.
V='1/2*k*(0.8*cos(x/2)-.4)^2-39.24*cos(x)'
dV=diff(V,'x')
dV900=subs(dV,'k',900)
theta900=solve(dV900,'x')
%%%%%%%%%%% end of script %%%%%%%%%
```

Running script 2 yields

```
V =

1/2*k*(0.8*cos(x/2)-.4)^2-39.24*cos(x)

dV =

-.4*k*(.8*cos(1/2*x)-.4)*sin(1/2*x)+39.24*sin(x)

dV900 =

-360*(.8*cos(1/2*x)-.4)*sin(1/2*x)+39.24*sin(x)

theta900 =

  6.2831853071795864769252867665590
                                  0.
 -1.6261025608933934751793403424078
  1.6261025608933934751793403424078
```

MATLAB has found four solutions. The second (theta = 0) is unstable. The fourth solution (theta = 1.6261) is the one corresponding to the relative minimum in the graph above.

Summary:

$k = 100$ N/m

    $\theta = 0$ Stable

$k = 900$ N/m

    $\theta = 0$ Unstable

    $\theta = 1.626$ rads (93.18°)   Stable